Lecture-Tutorials for Introductory Astronomy

Jeffrey P. Adams
Montana State University

Edward E. Prather
University of Arizona

Timothy F. Slater
University of Arizona

with contributions by

Jack A. Dostal
Montana State University

and the

Conceptual Astronomy and Physics Education Research (CAPER) Team

PRENTICE HALL SERIES IN EDUCATIONAL INNOVATION

PEARSON
Prentice
Hall

Upper Saddle River, New Jersey 07458

Associate Editor: Christian Botting
Senior Editor: Erik Fahlgren
Editor in Chief, Science: John Challice
Production Editor: Debra A. Wechsler
Executive Managing Editor: Kathleen Schiaparelli
Vice President of Production and Manufacturing: David W. Riccardi
Executive Marketing Manager: Mark Pfaltzgraff
Manufacturing Buyer: Alan Fischer
Manufacturing Manager: Trudy Pisciotti
Director of Creative Services: Paul Belfanti
Art Director: Jayne Conte
Cover Design: Bruce Kenselaar
Editorial Assistant: Andrew Sobel

© 2005 Pearson Education, Inc.
Pearson Prentice Hall
Pearson Education, Inc.
Upper Saddle River, NJ 07458

Printed in the United States of America
10 9 8 7 6 5 4

ISBN 0-13-147997-0

Pearson Education LTD., *London*
Pearson Education Australia PTY, Limited, *Sydney*
Pearson Education Singapore, Pte. Ltd
Pearson Education North Asia Ltd, *Hong Kong*
Pearson Education Canada, Ltd., *Toronto*
Pearson Educación de Mexico, S.A. de C.V.
Pearson Education -- Japan, *Tokyo*
Pearson Education Malaysia, Pte. Ltd

Table of Contents

SKIP

SKIP

SKIP

SKIP

SKIP?

No Doppler shift

© PEARSON PRENTICE HALL

LECTURE-TUTORIALS FOR INTRODUCTORY ASTRONOMY
FIRST EDITION

Table of Contents

Instructor's Preface

Each year, over 200,000 students take introductory astronomy—hereafter referred to as ASTRO 101; the majority of these students are non-science majors. Most are taking ASTRO 101 to fulfill a university science requirement and many approach science with some mix of fear and disinterest. The traditional approach to winning over these students has been to emphasize creative and engaging lectures, taking full advantage of both demonstrations and awe-inspiring astronomical images. However, what a growing body of evidence in astronomy and physics education research has been demonstrating is that even the most popular and engaging lectures do not engender the depth of learning for which faculty appropriately aim. Rigorous research into student learning tells us that one critical factor in promoting classroom learning is students' active "minds-on" participation. This is best expressed in the mantra: "It's not what the teacher does that matters; it's what the students do."

Lecture-Tutorials for Introductory Astronomy has been developed in response to the demand from astronomy instructors for easily implemented student activities for integration into existing course structures. Rather than asking faculty—and students—to convert to an entirely new course structure, our approach in developing *Lecture-Tutorials* was to create classroom-ready materials to augment more traditional lectures. Any of the activities in this manual can be inserted at the end of lecture presentations and, because of the education research program that led to the activities' development, we are confident in asserting that the activities will lead to deeper and more complete student understanding of the concepts addressed.[1]

Each *Lecture-Tutorial* presents a structured series of questions designed to confront and resolve student difficulties with a particular topic. Confronting difficulties often means answering questions incorrectly; this is expected. When this happens, the activities are crafted to help a student understand where her or his reasoning went wrong and to develop a more thorough understanding as a result. Therefore, while completing the activities, students are encouraged to focus more on their reasoning and less on trying to guess an expected answer. The activities are meant to be completed by students working in pairs who "talk out" the answers with each other to make their thinking explicit.

At the conclusion of each *Lecture-Tutorial*, instructors are strongly encouraged to engage their class in a brief discussion about the particularly difficult concepts in the activity—an essential implementation step that brings closure to the activity. The online *Instructor's Guide* also provides "post-tutorial" questions that can be used to gauge the effectiveness of the *Lecture-Tutorial* before moving on to new material.

Acknowledgments

Lecture-Tutorials for Introductory Astronomy was developed by the Conceptual Astronomy and Physics Education Research (CAPER) Team at Montana State University and the University of Arizona with generous support from the National Science Foundation (NSF CCLI #9952232 and NSF Geosciences Education #9907755). Gina Brissenden provided initial project definition, management, and graphical layout efforts. Jack Dostal and Larry Watson assisted directly in writing activities. Numerous other individuals contributed to this project through critical assessment and the national field-testing of the materials. These individuals include Ingrid Balsa, Chija Skala Bauer, Tom Brown, Dave Bruning, Beth

---

[1] Instructors can go to http://www.prenhall.com/tiponline for online *Instructor's Guide* that gives detailed information on classroom implementation as well as evidence of the efficacy of specific activities.

Hufnagel, Lauren Jones, Janet Landato, Ed Murphy, and Erika Offerdahl. Particularly noteworthy were the extensive reviews and suggestions provided by Janelle Bailey, Gina Brissenden and Steve Shawl, which continually kept us on our toes. In addition we must thank Christian Botting who helped us with day-to-day publication issues and Alison Reeves who repeatedly encouraged us to continue and helped us frame the initial ideas for this work. Most importantly, we wish to express our appreciation to our students who patiently endured early versions of these tutorials and unselfishly provided extensive feedback.

Note to the Student

Welcome to the study of astronomy! You are about to embark on a grand study of the cosmos. To help you better understand the topics of your course we have created this series of activities called *Lecture-Tutorials*. In each activity, you are asked a short series of questions that will require you to work in collaboration with your classmates to help you learn important and difficult concepts in astronomy. For every question in these activities it is important that you write out a detailed answer. This is critical because you will certainly be using these materials to study for exams. It is also important because part of the learning process is being able to express complex ideas in writing.

We strongly encourage you to actively engage in completing these activities in collaboration with another student. The process of deciphering the questions and negotiating a common language to write your answers will help you understand the concepts more deeply. Specifically, the *Lecture-Tutorials* are designed to give you a starting point to think carefully and talk with others about concepts in astronomy. Above all, have fun exploring astronomy!

J.P. Adams, E.E. Prather, T.F. Slater and the
Conceptual Astronomy and Physics
Education Research (C.A.P.E.R.) Team

In the celestial sphere model, Earth is stationary and the stars are carried on a sphere that rotates about an axis through the North Star. In Figure 1 below, two stars, A and B, are each shown at four different positions (1, 2, 3, and 4) through which each star will pass during the course of one revolution of the celestial sphere. In addition, your location in the northern hemisphere and the portion of the celestial sphere that is above your horizon are shown.

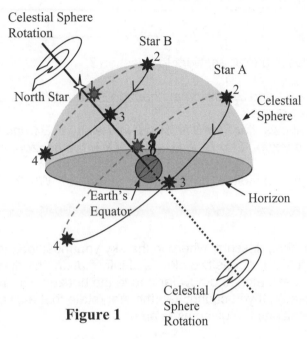

Figure 1

1) Is the horizon shown a real physical horizon or an imaginary plane that extends from the observer and Earth out to the stars?

2) Can the observer shown see a star when it is located below the horizon? Why or why not?

3) Is either Star A or B ever in an unobservable position? If so, which position(s)?

4) When a star travels from a position below the observer's horizon to a position above the horizon, is that star rising or setting?

5) When a star travels from a position above the observer's horizon to a position below the horizon, is that star rising or setting?

LECTURE-TUTORIALS FOR INTRODUCTORY ASTRONOMY
                                                         FIRST EDITION

about here, tell them about North,
[Many think South is down, or through E opposite of N. ☆]

6) Star A is just visible above your eastern horizon at position 1. At which of the numbered positions is it just visible above your western horizon?

7) At which position(s), if any, does Star B rise and set?

    4

8) Two students are discussing their answers to question 7.

    Student 1: *Locations B1 and B3 are on my horizon because they are rising and setting just like A1 and A3.*
    Student 2: *Figure 1 shows that Star B is lowest in the sky at B4 and is just above the northern horizon. Star B never goes below the horizon.*

    Do you agree or disagree with either or both of the students? Why?

9) For each indicated position, describe where in the sky you must look to see the star at that time. Each description requires two pieces of information: the direction you must face (north, northeast, east, etc.) and how far above the horizon you must look (low, high, or directly overhead). If you cannot see the star, state that explicitly. The descriptions for four positions are given as examples.

    a) A1: *east, low*

    b) A2:

    c) A3:

    d) A4:

    e) North Star: *north, high*

    f) B1:

    g) B2: *directly overhead*

    h) B3: *northwest, high*

    i) B4:

    Check your answers with a nearby group and resolve any inconsistencies.

10) Does Star B ever set?

Part I: Looking North

For the activity, imagine that you are the observer shown in the northern hemisphere and that it is 6 PM. Looking north, the sky will appear as shown in Figure 1. The positions and motions of the stars in Figure 1 can be understood by imagining yourself as the observer at the center of the celestial sphere as shown in Figure 2. In the celestial sphere model, Earth is stationary and the stars are carried on a sphere that rotates about an axis through the North Star. Note that only the portion of the celestial sphere that is above your horizon is shown.

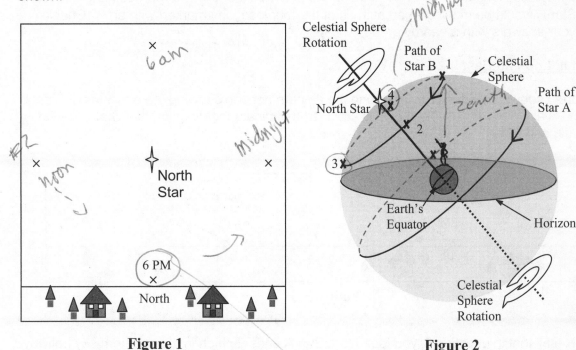

Figure 1   **Figure 2**

The ×'s in both figures represent four of the positions through which Star B will pass during the course of one revolution of the celestial sphere. Ignore Star A until question 5.

1) Note in Figure 1 the position of Star B at 6 PM. Circle the numbered position (1, 2, 3, or 4) in Figure 2 that corresponds to the location of Star B at 6 PM.

2) The rotation of the celestial sphere carries Star B around so that it returns to the same position at about 6 PM the next evening. Label each of the ×'s in both figures with the approximate time at which Star B will arrive (e.g., the location you circled in question 1 will be labeled "6 PM").

3) Using Figure 2, describe the direction you have to look to see Star B at 6 AM.

4) The position directly overhead is called the **zenith**. Label the direction of the zenith on Figure 2. How does the direction of the zenith compare to the direction that you identified in question 3?

5) Using Figure 2, describe in words the position of Star A halfway between rising and setting.

6) In Figure 1, use a dotted line to draw the entire path that Star B takes over the course of 24 hours. Next, draw an arrowhead on the path you just drew to represent the direction Star B would be moving when at each of the four locations marked with an ×. Check your answers with a nearby group. *Isn't it done already?*

Part II: Looking East

Figure 3 shows an extended view along the eastern horizon showing the positions of Stars A and B at 6 PM. The arrow shown is provided to indicate the direction that Star B will be moving at 6 PM.

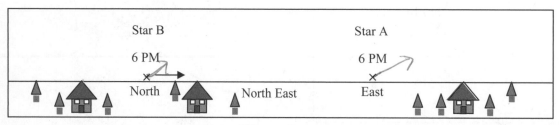

Figure 3

7) Recall that in question 5, you found that Star A ends up high in the southern sky halfway between rising and setting (and therefore never passes through your zenith). Draw a straight arrow at the × in the east in Figure 3 (the position of Star A at 6 PM) to indicate the direction Star A moves as it rises. Studying Figure 2 can also help clarify your answer.

8) Two students are discussing the direction of motion of a star rising directly in the east.

Student 1: *Stars move east to west so any star rising directly in the east must be moving straight up so that it can end up in the west. If the arrow were angled, the star would not set in the west.*

Student 2: *I disagree. From Figure 2, the path of Star A starts in the east, swings through the southern sky yet still sets in the west. To do this it has to move toward the south as it rises so I drew my arrow angled to the right.*

Do you agree or disagree with either or both of the students? Why?

9) Imagine you could see Star B at noon. Fifteen minutes later, in what direction will Star B have moved? Explain your reasoning.

N ☓ Down

10) Consider the student comment below.

Student: *The amount of time that all stars are visible above the horizon is 12 hours because it takes 12 hours for a star to rise in the east and then set in the west.*

Do you agree or disagree with this student statement? Why?

Consider the situation shown below in which the Sun and a group of constellations are shown at sunrise, Figure 4, and then shown again 8 hours later, Figure 5.

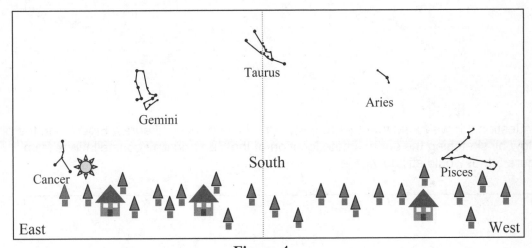

Figure 4

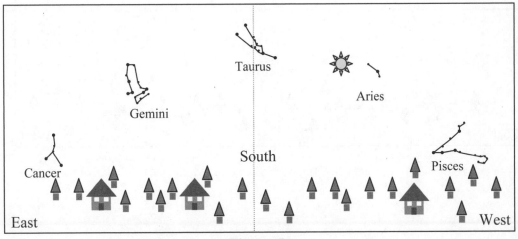

Figure 5

11) Consider the following debate between two students regarding the motion of the Sun and constellations shown in Figures 4 and 5.

Student 1: *We know the Sun rises in the east and moves through the southern part of the sky and then sets in the west. Eight hours after sunrise, it makes sense that the Sun will have moved from being on the eastern horizon near the constellation Cancer to being located high in the southwestern sky near the constellation Aries.*

Student 2: *You're forgetting that the stars and constellations also move from the east through the southern sky and to the west just like the Sun. So, the Sun will still be near Cancer eight hours later. So Figure 5 is drawn incorrectly. It should show that the constellations have all moved like the Sun, so Cancer would also be located high in the southwestern sky, with the Sun, eight hours later.*

Do you agree or disagree with either or both of the students? Why? Check your answers with a nearby group.

12) In question 11, we found that Figure 5 was drawn incorrectly. Redraw Figure 5 in the box below by sketching the approximate location of the Sun and any constellations from Figure 5 that would still be visible.

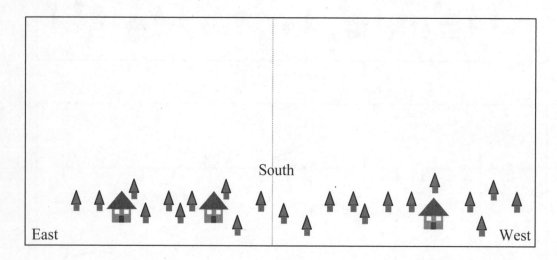

Part I: Monthly Differences

Figure 1 shows a sun-centered, or heliocentric, perspective view of the Earth-Sun system indicating the direction of both the daily rotation of the Earth about its own axis and its annual orbit about the Sun. You are the observer shown in Figure 1, located in the northern hemisphere while facing south.

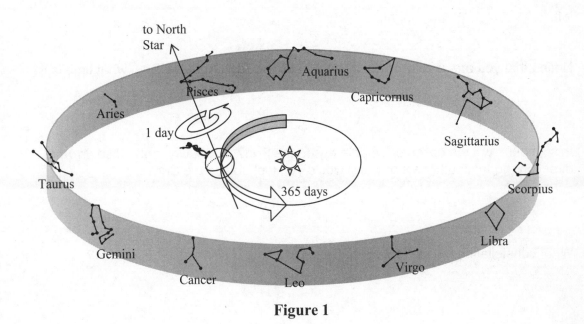

Figure 1

Figure 2 shows a horizon view of what you would see when facing south on this night at the same time as shown in Figure 1.

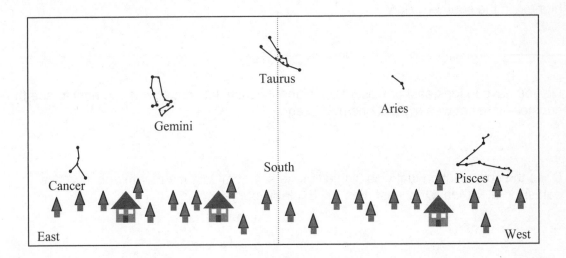

Figure 2

1) Which labeled constellation do you see highest in the southern sky?

2) What constellation is just to the left (i.e., east) and what constellation is just to the right (i.e., west) of the highest constellation at this instant?

   left: right:

3) Noting that you are exactly on the opposite side of Earth from the Sun, what time is it?

4) In six hours, will the observer be able to see the Sun? If not, why not? If so, in what direction (north, south, east or west) would you look to see the Sun?

5) What constellation will be behind the Sun at the time described in question 4?

6) When it is noon for the observer, what constellation will be behind the Sun?

7) One month later, the Earth will have moved one-twelfth of the way around the Sun. You are again facing south while observing at midnight. Which constellation will now be highest in the southern sky?

8) Do you have to look east or west of the highest constellation that you see now to see the constellation that was highest one month ago?

9) Does the constellation that was highest in the sky at midnight a month ago now rise earlier or later than it rose last month? Explain your reasoning.

Part II: Daily Differences

Figure 3 shows the same Earth-Sun view as before and the bright star Betelgeuse, which is between Taurus and Gemini.

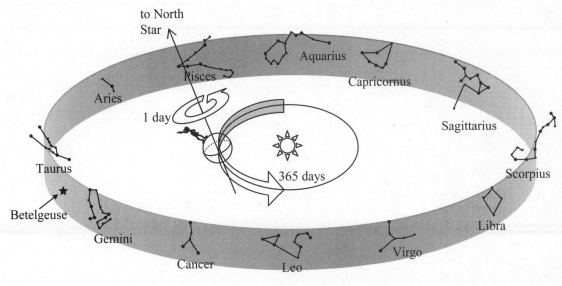

Figure 3

10) Last night you saw the star Betelgeuse exactly on your eastern horizon at 5:47 PM. At 5:47 tonight, will Betelgeuse be above, below, or exactly on your eastern horizon?

11) Two students are discussing their answers to question 10.

   Student 1:   *The Earth makes one complete rotation about its axis each day so Betelgeuse will rise at the same time every night. It will therefore be exactly on the eastern horizon.*

   Student 2:   *No. Because Earth goes around the Sun, the constellation Taurus rises earlier each month and so it must rise a little bit earlier each night. Betelgeuse must do the same thing. Tonight it would rise a little before 5:47 and be above the eastern horizon by 5:47.*

   Do you agree or disagree with either or both of the students? Why?

Part I: Solar Day

Figure 1 shows a top-down view of the Earth-Sun system. Arrows indicate the directions of the rotational and orbital motions of Earth. For the observer shown, the Sun is highest in the sky at 12 noon.

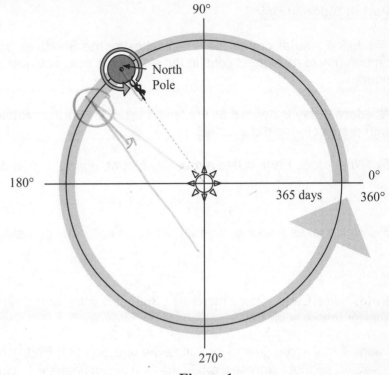

Figure 1

1) Earth orbits the Sun in a counterclockwise direction once every 365 days. Approximately how many degrees does Earth move along its orbit in one day?

2) As Earth orbits the Sun, it also rotates in a counterclockwise direction about its axis as shown in Figure 1. We define 24 hours as the time from when the Sun is highest in the sky one day to when it is highest in the sky the next day. How many degrees does Earth rotate about its axis in exactly 24 hours: 360°; slightly less than 360°; or slightly more than 360°?

3) How long does it take Earth to rotate exactly 360°: slightly less than 24 hours; 24 hours; or slightly more than 24 hours?

4) Two students are discussing their answer to questions 2 and 3.

    Student 1: *Earth rotates about its axis once every 24 hours and one rotation equals 360°.*

    Student 2: *No. When Earth has gone around 360° it has also moved a small amount counterclockwise around the Sun, which means the Sun is not yet at its highest point. Earth must spin a little bit more for the Sun to reach its highest point.*

    Do you agree or disagree with either or both of the students? Why?

Part II: Sidereal Day

We define a **solar day** as the time it takes for the Sun to go from its highest point in the sky on one day to its highest point in the sky on the next day and we divide that time into 24 hours.

A **sidereal day** is defined as the time it takes for Earth to rotate *exactly* 360° about its axis with respect to the distant stars.

5) When does Earth rotate a greater amount, during a solar day or during a sidereal day?

6) Which takes a shorter amount of time, a solar day or a sidereal day?

Note: Since Earth rotates more than 360° in a solar day, a sidereal day is about 4 minutes shorter than a solar day.

Imagine that some time in the future the direction that Earth orbits the Sun is somehow reversed so that Earth now orbits the Sun approximately 1° *clockwise* each day. However, the rotation about its own axis remains counterclockwise at the same rate.

7) In the space below, create a sketch similar to Figure 1 to depict this imaginary situation.

8) Through how many degrees will Earth now rotate in a <u>sidereal</u> day?

9) Through how many degrees will Earth now rotate in a <u>solar</u> day?

10) Which is now longer, the solar or the sidereal day?

11) Is a sidereal day now longer, shorter, or the same length as a sidereal day was before we changed Earth's orbital direction?

12) Is a solar day now longer, shorter, or the same length as a solar day was before we changed Earth's orbital direction?

For all parts of this activity, it is helpful to imagine that the stars are so bright (or our Sun so dim) that the stars can be seen during the day so that your sky might appear as in Figure 1.

Part I: Daily Motion

On December 1, at noon, you are looking toward the south and see the Sun among the stars of the constellation Scorpius as shown in Figure 1.

1) At 3 PM that afternoon, will the Sun appear among the stars of the constellation Sagittarius, Scorpius, or Libra?

2) Two students are discussing their answers to question 1.

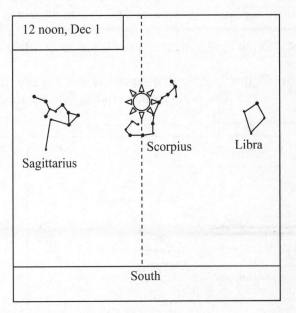

Figure 1

> **Student 1:** *The Sun moves from the east through the southern part of the sky and then to the west. By 3 PM it will have moved from being high in the southern sky to the west into the constellation Libra.*

> **Student 2:** *You're forgetting that stars and constellations will rise in the east, move through the southern sky and then set in the west just like the Sun. So the Sun will still be in Scorpius at 3 PM.*

Do you agree or disagree with either or both of the students? Why?

Recall that in the celestial sphere model, the stars' daily motions result from the rotation of the celestial sphere.

3) Is it reasonable to account for the Sun's **daily motion** by assuming that the Sun is at a fixed position on the celestial sphere (in this case in the location of the constellation Scorpius) and is carried along its path across the sky by the sphere's rotation? Explain why or why not.

Part II: Monthly Changes

By careful observation of the Sun's position in the sky throughout the year, we find that the celestial sphere rotates slightly more than 360° every 24 hours. Figure 2 shows the same view of the sky (as Figure 1) but on December 2 at noon. For comparison, the view from the previous day at the same time is also shown in gray.

4) Draw the location of the Sun as accurately as possible in Figure 2.

5) Figure 3 shows the same view of the sky (as Figure 1) one month later on January 1 at noon. Draw the location of the Sun as accurately as possible in this figure.

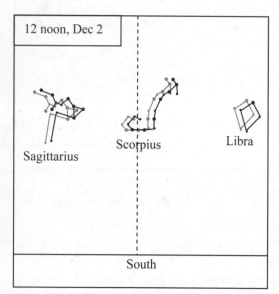
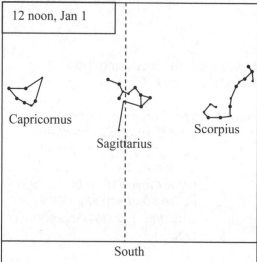

Figure 2   Figure 3

6) Two students are discussing their answers to questions 4 and 5.

  Student 1: *The Sun will always lie along the dotted line in the figures when it's noon.*

  Student 2: *But, we saw in question 3 that the Sun's motion can be modeled by assuming it is stuck to the celestial sphere. The Sun must, therefore, stay in Scorpius.*

  Student 1: *If that were true then by March the Sun would be setting at noon. The Sun must shift a little along the celestial sphere each day so that in 30 days it has moved to the east into the next constellation.*

Do you agree or disagree with either or both of the students? Why?

7) Why is it reasonable to think of the Sun as attached to the celestial sphere over the course of a single day as suggested in question 3 even though we know from questions 5 and 6 that the Sun's position is not truly fixed on the celestial sphere?

Part III: The Ecliptic

The zodiacal constellations were of special interest to ancient astronomers because these are the constellations through which the Sun moves throughout the year. This was incorporated into their celestial sphere model by having the Sun loosely fixed to the celestial sphere but allowing it to slip a small amount each day. The Sun's position on the celestial sphere (among the stars in the constellation Scorpius) on December 1 is shown in Figure 4.

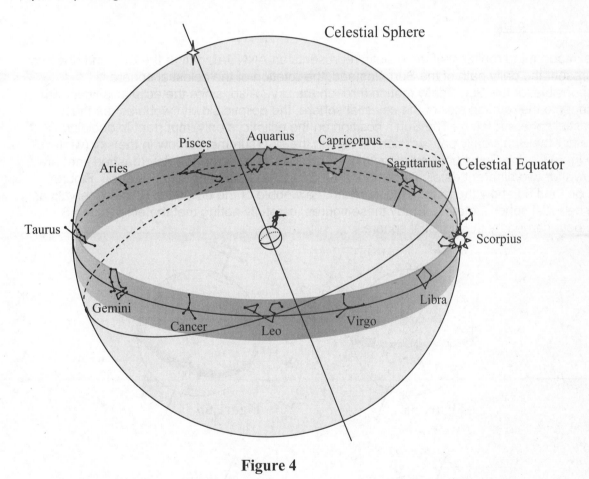

Figure 4

8) Draw where the Sun will be located on the celestial sphere on January 1. Label this position "Jan 1".

9) For the other constellations, draw in a Sun and label the constellation with the approximate date that the Sun will be located there.

The line drawn through these constellations, tracing out the Sun's annual path, is called the **ecliptic**.

10) Label the ecliptic in Figure 4.

11) About how many times does the celestial sphere rotate in the time it takes the Sun to move between two adjacent constellations (i.e., 1/12 of the way around) along the ecliptic?

12) How long does it take the Sun to make one complete trip around the ecliptic (i.e., from Scorpius to Scorpius)?

Part IV: Wrap Up

It is important to realize that the ecliptic represents an ANNUAL drift of the Sun and does not represent the daily path of the Sun. Instead, the rotation of the celestial sphere is responsible for the Sun's daily motion through the sky. Also, since the ecliptic is tilted with respect to the rotation axis of the celestial sphere, the ecliptic slowly "wobbles" as the celestial sphere rotates. The Sun's position on the ecliptic is only important in deciding whether the Sun's daily path will carry it high in the sky (summer) or low in the sky (winter). In Figure 5a, the Sun's position along the ecliptic and its path for one day (dashed line) are shown for two different dates: December 1 (in Scorpius), and June 1 (in Taurus). Figures 5b, 5c, and 5d show the path of the Sun and the wobble of the ecliptic at 6-hour intervals as the celestial sphere rotates. Study these figures, carefully noting that the ecliptic and Sun are both carried by the celestial sphere.

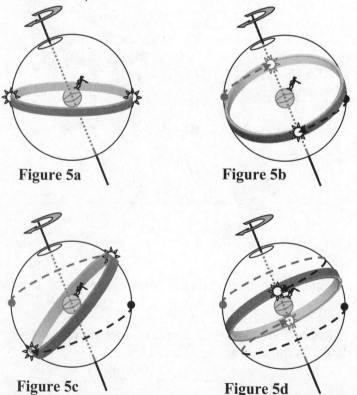

Figure 5a   **Figure 5b**

Figure 5c   **Figure 5d**

13) On Figure 5d, label the ecliptic (Sun's annual path) and the Sun's daily path for December 1 and June 1.

14) Which Figure (5a, 5b, 5c, or 5d) shows the Sun at noon, low in the southern sky, when it would be among the stars of the constellation Scorpius?

15) Which Figure (5a, 5b, 5c, or 5d) shows the Sun at noon, high in the southern sky, when it would be among the stars of the constellation Taurus?

Figure1 illustrates the sky as seen from the continental United States (US). It shows that the Sun's daily path across the sky (dashed line) is longest on June 21 and shortest on December 21. In addition, on June 21, which is called the summer solstice, the Sun reaches its maximum height in the southern sky above the horizon at about noon. The figure shows that the Sun never actually reaches the zenith for any observer in the continental US. In other words, the Sun is never directly overhead. Over the six months following the summer solstice, the height of the Sun at noontime moves progressively lower and lower until December 21, the winter solstice. Thus, we see that the path of the Sun through the southern sky changes considerably over the course of a year.

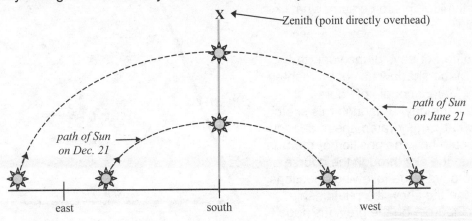

Figure 1

1) According to Figure 1, in which direction would you look to see the Sun when it reaches its highest position in the sky today?

    Circle one: east   southeast   <u>south</u>   southwest   west

2) If it is wintertime right now (just after the winter solstice), how does the height of the Sun at noon change as summer approaches?

    Circle one: <u>increases</u>   stays the same   decreases

3) If Figure 1 is a reasonable representation for observers in the continental US, is there ever a time of year when the Sun is directly overhead at the zenith (looking straight up) at noon? If so, on what date does this occur?   No

Give answers

4) During what time(s) of year would the Sun rise:
   a) north of east?

   b) south of east?

   c) directly in the east?

5) Does the Sun always set in precisely the same location throughout the year? If not, describe how the location of sunset changes throughout the year.

Figure 2 shows a small, vertical stick, which casts a shadow while it rests on a large piece of paper or poster-board. You can think of this to be somewhat like a sundial.

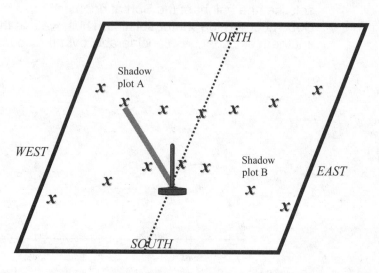

Figure 2

For two different days of the year, the very top of the shadow has been marked with an "×" every couple of hours throughout the day. Although this sketch is somewhat exaggerated, these *shadow plots* indicate how the position of the Sun changes in the sky through the course of these two days. The following questions are designed to show the relationship between Figure 1 on the previous page and Figure 2 at right.

6) Using Figures 1 and 2, in what direction would the shadow of the Sun be cast on the poster-board if the Sun rises in the southeast?

   Circle one: west   northwest   north   northeast   east   southeast

7) Clearly circle the × for the shadow that corresponds to the time of noon for plot A and for plot B.

8) Compare the position of the × that corresponds to noon for shadow plots A and B. Which shadow plot (A or B) corresponds to a path of the Sun in which the Sun is highest in the sky at noon? Explain your reasoning.

9) Which shadow plot (A or B) most closely corresponds to the Sun's path through the sky during the summer and which corresponds with the winter? Explain your reasoning.

10) On Figure 2, sketch the Sun's position at sunrise in the summer and label the × that the Sun's shadow would make at this time.

11) Based on the shadow plots in Figure 2, during which time of the year (summer or winter) does the Sun rise south of east? Explain your reasoning.

12) If shadow plot A corresponds to the path of the Sun on the day of the winter solstice, is it possible that there would ever be a time when the stick would cast a shadow longer than the one shown along the north-to-south line that indicates the Sun's position at noon? Explain your reasoning.

13) If shadow plot B corresponds to the path of the Sun on the day of the summer solstice, is it possible that there would ever be a time when the stick would cast a shadow shorter than the one shown along the north-to-south line that indicates the Sun's position at noon? Explain your reasoning.

14) If you were to mark the top of the stick's shadow with an ×, where would the × be placed along the north-to-south line to indicate the Sun's position at noon *today*? Clearly explain why you placed the × where you did.

15) Will the stick ever cast a shadow along the north-to-south line that extends to the south of the stick? Explain your reasoning.

16) Is there ever a clear (no clouds) day of the year in the continental US when the stick casts no shadow? If so, when does this occur and where exactly in the sky does the Sun have to be?

Consider the star map for July at midnight shown in Figure 1. In particular, notice that the directions of north and east have been identified and that the names of different star groups (constellations) have been provided.

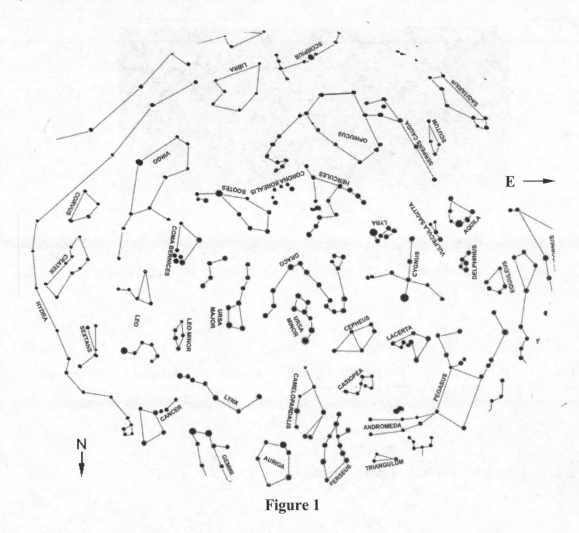

Figure 1

1) Which star group will appear highest in the night sky at this particular time?

2) Figure 2 shows a south-facing horizon view star map for July at midnight. What is the name of the star group that appears highest in the sky on this south-facing horizon view star map? (Hint: refer to the names provided in Figure 1.)

Figure 2

3) How would you have to hold, rotate, fold, and/or change the overhead view star map shown in Figure 1 so that it could be used as a south-facing star map like the one provided in Figure 2?

4) How would your answer to the previous question change if you wanted to use the star map from Figure 1 as a north-facing map?

5) Do you still agree with your answer to question 1? Why or why not?

6) When looking at the overhead view star map from Figure 1:

a) on what part of the map (left, right, top, bottom, or center) is the star group that will appear highest in the night sky? What is the name of this star group?

b) on what part of the map (left, right, top, bottom, or center) is the star group that will appear near the southern horizon? What is the name of this star group?

c) on what part of the map (left, right, top, bottom, or center) is the star group that will appear near the eastern horizon? What is the name of this star group?

10 min

see p.27

Figure 1 shows Earth, the Sun, and five different possible positions for the Moon during one full orbit (dotted line). It is important to recall that one half of the Moon's surface is illuminated by sunlight at all times. For each of the five positions of the Moon shown below, the Moon has been shaded on one side to indicate the half of the Moon's surface that is **NOT** being illuminated by sunlight. Note that this drawing is not to scale.

1) Which Moon position (A – E) best corresponds with the moon phase shown in the upper right corner of Figure 1? Make sure that the moon position you choose correctly predicts a moon phase in which only a small crescent of light on the left-hand side of the Moon is visible from Earth.

Enter the letter of your choice:___ *D*

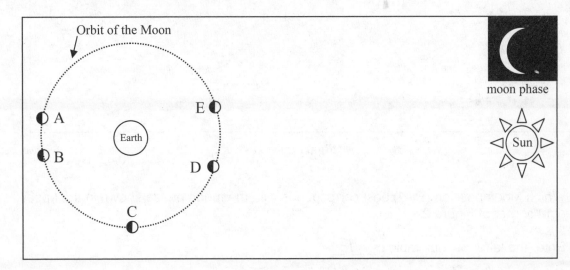

Figure 1

2) In the blank boxes below, sketch how the Moon would appear from Earth for the four Moon positions that you did **not** choose in question 1. Be sure to label each sketch with the corresponding letter indicating the Moon's position from Figure 1.

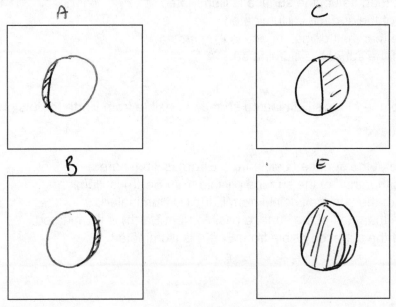

3) Shade in each of the four Moons drawn in Figure 2 to indicate which portion of the Moon's surface will **NOT** be illuminated by sunlight. Use Figure 2 to answer questions 4 – 7.

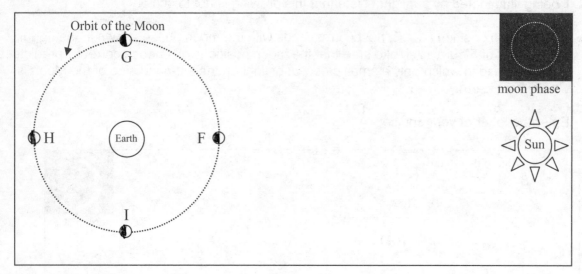

Figure 2

4) Which Moon position (F – I) best corresponds with the moon phase shown in the upper right corner of Figure 2?

Enter the letter of your choice: _F_

5) How much of the entire Moon's surface is illuminated by the Sun during this phase?

Circle one:
- ⊗ None of the surface is illuminated.
- Less than half of the surface is illuminated.
- ◉ Half of the surface is illuminated.
- More than half of the surface is illuminated.
- All of the surface is illuminated.

6) How much of the Moon's illuminated surface is visible from Earth for this phase of the Moon?

Circle one:
- ◉ None of the surface (visible from Earth) is illuminated.
- Less than half of the surface (visible from Earth) is illuminated.
- Half of the surface (visible from Earth) is illuminated.
- More than half of the surface (visible from Earth) is illuminated.
- All of the surface (visible from Earth) is illuminated.

at Point
6

7) Would your answers to questions 5 and 6 be the same if the Moon were in the third quarter phase? Explain your reasoning.

always half is illuminated

No 5 - same
6 - No

you only see $\frac{1}{4}$ at 3rd quarter ← see half the visible part

8) Consider the following debate between two students about the cause of the phases of the Moon.

Student 1: *The phase of the Moon depends on how the Moon, Sun and Earth are aligned with one another. During some alignments only a small portion of the Moon's surface will receive light from the Sun, in which case we would see a crescent moon.*

Student 2: *I disagree. The Moon would always get the same amount of sunlight; it's just that in some alignments Earth casts a larger shadow on the Moon. That's why the Moon isn't always a full moon.*

Do you agree or disagree with either or both of the students? Why?

Disagree with both

1) If the Moon is a full moon tonight, will the Moon be waxing or waning one week later? Which side of the Moon (right or left) will appear illuminated at this time?

   Circle one: Waxing or <u>Waning</u>

   Circle one: Right or <u>Left</u>

2) Where (in the southern sky, on the eastern horizon, on the western horizon, high in the sky, etc.) would you look to see the full moon when it starts to rise? What time of day would this happen? E 6pm Sunset

3) Where (in the southern sky, on the eastern horizon, on the western horizon, high in the sky, etc.) would you look to see the Sun when the full moon starts to rise?

   W horizon

4) Where (in the southern sky, on the eastern horizon, on the western horizon, high in the sky, etc.) would you look to see the new moon, if it were visible, when it starts to rise? What time of day would this happen? E horiz, Sunrise

5) If the Moon is a full moon when it rises in the evening, which of the phases shown below (A – H) will it be in when it sets?

   Letter of moon phase: ___D___

Figure 1 shows the position of the Sun, Earth and Moon for a particular phase of the Moon. The Moon has been shaded on one side to indicate the portion of the Moon that is **NOT** being illuminated by sunlight. A stick-figure person has been placed on Earth to indicate an observer's position at noon. Recall that with this representation Earth will complete one counterclockwise rotation in each day. Note that this drawing is not to scale.

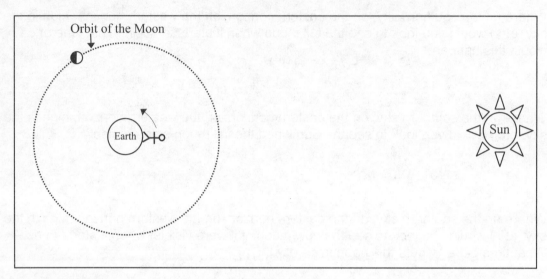

Figure 1

6) What time of day is it for the person shown in Figure 1?

   Circle one: 6am (sunrise)   12pm (noon)   6pm (sunset)   12am (midnight)

7) Draw a stick-figure person on Earth in Figure 1 for each of the three times that you did not choose in question 6. Label each of the stick-figures that you drew with the time that the person would be located there.

8) Answer the following questions for the position of the Moon shown in Figure 1.

   a) Which moon phase would an Earth observer see?

   b) At what time will the Moon shown appear highest in the sky?

   c) At what time will the Moon shown appear to rise?

   d) At what time will the Moon shown appear to set?

9) At what time would you look to see a first quarter Moon at its highest position in the sky?

10) If the Sun set below your western horizon about 2 hours ago, and the Moon is barely visible on the eastern horizon, what phase would the Moon be in at this time and location?

11) A friend comments to you that there was a beautiful, thin sliver of a Moon visible in the early morning just before sunrise. Which phase of the Moon would this be, and in what direction would you look to see the Moon (in the southern sky, on the eastern horizon, on the western horizon, high in the sky, etc.)?

Part I: Luminosity, Temperature and Size

Imagine you are comparing the abilities of electric hot plates of different sizes and temperatures to fully cook two identical large pots of spaghetti. Note that the pots are all as large as the largest hot plate. When a hot plate is at one of the temperature settings (low, med, high), the hot plate is depicted as a shade of gray as shown in question 1. The lighter the shade of gray, the higher the temperature setting of the hot plate.

1) For each pair of hot plates shown below, circle the one that will cook the large pot of spaghetti more quickly. If there is no way to tell, state that explicitly.

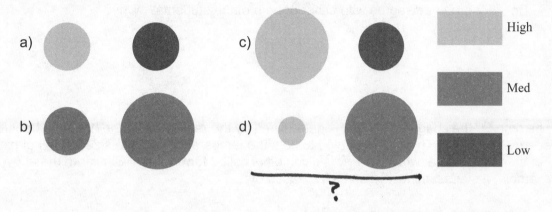

2) If you use two hot plates of the same size, can you assume that the hot plate that can cook a large pot of spaghetti first is at the higher temperature? Which lettered example above supports your answer?

3) If you use two hot plates at the same temperature, can you assume that the hot plate that can cook a large pot of spaghetti first is larger? Which lettered example above supports your answer?

4) If you use two hot plates of different sizes, can you assume that the hot plate that can cook a large pot of spaghetti first is at a higher temperature? Which lettered example above supports your answer?

5) Two students are discussing their answers to question 4:

Student 1:   *In 1d, the hot plate on the left cooks the spaghetti quicker than the one on the right even though it is smaller. The hot plate's higher temperature is what makes it cook the spaghetti more quickly.*

Student 2:   *But the size of the hot plate also plays a part in making it cook fast. If the hot plate on the left were the size of a penny, the spaghetti would take a really long time to cook. I bet that if the size difference were great enough, the one at the lower temperature could cook the spaghetti first.*

Do you agree or disagree with either or both of the students? Why?

The time for the spaghetti to cook is determined by the rate at which the hot plate transfers energy to the pot. This rate is related to both the temperature and the size of the hot plate. For stars, the rate at which energy is given off is called **luminosity**. Similar to the above example, a star's luminosity can be increased by:

- increasing its temperature; and/or

- increasing its surface area (or size).

This relationship between luminosity, temperature and size allows us to make comparisons between stars.

6) If two hot plates have the same temperature and one cooks the pot of spaghetti more quickly, what can you conclude about the sizes of the hot plates?

7) Likewise, if two stars have the same surface temperature and one is more luminous, what can you conclude about the sizes of the stars?

8) If two stars have the same surface temperature and are the same size, which star, if either, is more luminous? Explain your answer.

9) If two stars are the same size but one has a higher surface temperature, which star, if either, is more luminous? Explain your answer.

Part II: Application to the H-R Diagram

The graph below plots the luminosity of a star on the vertical axis against the star's surface temperature on the horizontal axis. This type of graph is called an H-R diagram. Use the H-R diagram below and the relationship between a star's luminosity, temperature and size (as described on the previous page) to answer the following questions concerning the stars labeled **s – y**.

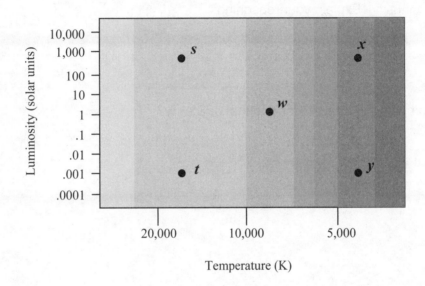

10) Stars **s** and **t** have the same surface temperature. Given that Star **s** is actually much more luminous than Star **t**, what can you conclude about the size of Star **s** compared to Star **t**? Explain your answer.

11) Star **s** has a greater surface temperature than Star **x**. Given that Star **x** is actually just as luminous as Star **s**, what can you conclude about the size of Star **x** compared to Star **s**? Explain your answer.

12) Based on the information presented in the H-R diagram, which star is larger, *x* or *y*? Explain.

13) Based on the information presented in the H-R diagram, which star is larger, *y* or *t*? Explain.

14) On the H-R diagram, draw a "**z**" at the position of a star smaller in size than Star *w* but with the same luminosity. Explain your reasoning

15) Why can't you compare the size of Star *s* to that of Star *w*?

Part I: Spectral Curves

White light is made up of all colors of light. We can see the individual colors when we shine white light through a prism or look at a rainbow. The composition of light—called a spectrum—can be represented by a spectral curve, which shows the brightness of each color (or wavelength). For a specific color of light on the horizontal axis, an object's brightness (or, more correctly, energy output per second) is represented by the height of the curve at that point. Figure 1 shows a spectral curve for a source emitting more red and orange light than indigo and violet. Notice that the red end of the curve is higher than the violet end so the object will appear slightly reddish in color.

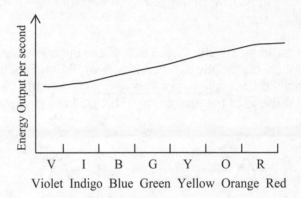

1) Which color of light is most intense in Figure 1?

2) Imagine that the blue light and orange light from the source were blocked. What color(s) would now be present in the spectrum of light observed?

Figure 1

3) Which of the following is the most accurate spectral curve for the spectrum described in question 2?

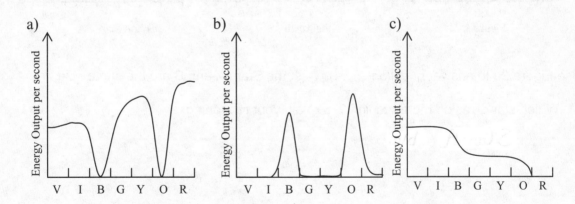

4) What colors of light are present in 3b above?

5) What colors are present in 3c above? Would this object appear reddish or bluish?

Part II: Blackbody Curves

Different colors of light are manifestations of the same phenomenon but have different wavelengths. For example, red light has a wavelength between 650 nm and 750 nm, while violet light has a shorter wavelength between 350 nm and 450 nm. Stars also give off light at wavelengths outside the visible part of the spectrum as seen in Figures 2a, 2b, and 2c.

The two most important features of a star's blackbody curve are:
- its maximum height or peak — an indication of the star's energy output; and
- the wavelength at which this peak occurs (called the peak wavelength) — an indication of the star's temperature (the longer the peak wavelength, the cooler the star).

For example, if Star A and Star B are the same size and temperature, they will have identical blackbody curves. However, if Star B is the same size as Star A, but is cooler, its energy output is <u>less at all wavelengths</u> and the peak occurs <u>at a longer wavelength</u> (toward the red end of the spectrum). This is shown in Figure 2a.

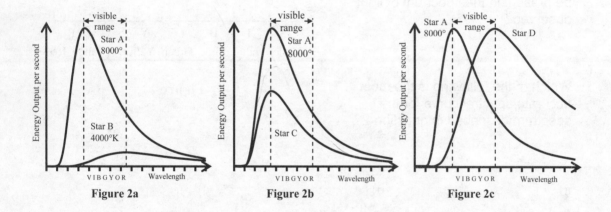

Figure 2a   Figure 2b   Figure 2c

Use Figure 2a to answer questions 6 – 9. Assume Stars A and B are the same size.

6) Which star gives off more red light? Explain your reasoning.

7) Which star gives off more blue light? Explain your reasoning.

8) Which star looks redder? Explain your reasoning.

9) Two students are discussing their answers to question 8.

Student 1: *Star A looks redder because it is giving off more red light than Star B.*

Student 2: *I disagree, you're ignoring how much blue light Star A gives off. Star A gives off more blue light than red light so it looks bluish. Star B gives off more red than blue so it looks reddish. That's why Star B looks redder than Star A.*

Do you agree or disagree with either or both of the students? Why?

10) Using the blackbody curves for the stars shown in Figure 2b, circle the correct answer for each characteristic of the curves below.

Longer peak wavelength	Star A	Star C	Same •	
Lower surface temperature	Star A	Star C	Same •	
Looks red	Star A	Star C	Both	Neither •
Looks blue	Star A	Star C	Both •	Neither
Greater energy output	Star A •	Star C		

11) How must Star C be different from Star A to account for the difference in energy output?

12) Two students are discussing their answers to question 11.

Student 1: *The peaks are at the same place so they must be at the same temperature. If Star C were as big as Star A, it would have the same output. Since the output is lower, Star C must be smaller.*

Student 2: *No. If its output is lower, it must be cooler. Since the temperature of the two stars are the same, they must be the same size.*

Do you agree or disagree with either or both of the students? Why?

Consider the blackbody curves for the stars shown in Figure 2c when answering questions 13 – 15.

13) For each star, describe its color as either reddish or bluish.

    Star A: Star D:

14) Which star has the greater surface temperature? Explain your reasoning.

15) Which star is larger? Explain your reasoning. (Hint: consider how the energy output and temperatures for the two stars compare.)

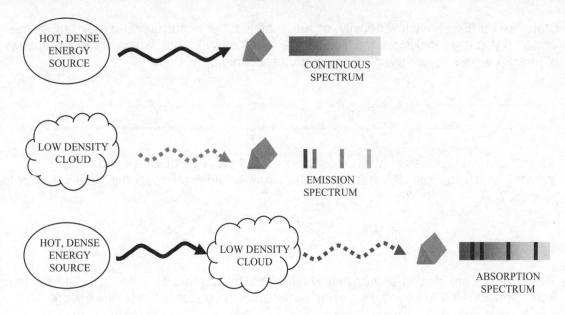

1) What type of spectrum is produced when the light emitted directly from a hot, dense object passes through a prism?

2) What type of spectrum is produced when the light emitted directly from a low density cloud passes through a prism?

3) Describe in detail the source of light and the path the light must take to produce an absorption spectrum.

4) There are dark lines in the absorption spectum that represent missing light. What happened to this light that is missing from the absorption line spectrum?

5) Stars like our Sun have low density, gaseous atmospheres surrounding their hot, dense cores. If you were looking at the Sun's (or any star's) spectrum, which of the three types of spectra would be observed? Explain your reasoning.

6) If a star existed that was only a hot dense core and did NOT have a low density atmosphere surrounding it, what type of spectrum would you expect this particular star to produce?

7) Two students are looking at a brightly lit full moon, illuminated by reflected light from the Sun. Consider the following discussion between the two students about what the spectrum of moonlight would look like.

   Student 1: *I think moonlight is just reflected sunlight, so we will see the Sun's absorption line spectrum.*
   Student 2: *I disagree, an absorption spectrum has to come from a hot dense object. Since the Moon is not a hot dense object it can't give off an absorption line spectrum.*

   Do you agree or disagree with either or both of the students? Why?

8) Imagine that you are looking at two different spectra of the Sun. Spectrum #1 is obtained using a telescope that is in a high orbit far above Earth's atmosphere. Spectrum #2 is obtained using a telescope located on the surface of Earth. Label each spectrum below as either Spectrum #1 or Spectrum #2.

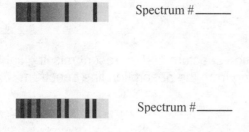

   Spectrum #_____

   Spectrum #_____

   Explain the reasoning behind your choices.

The absorption line spectra for six hypothetical stars, each with different temperatures, are shown below. For each absorption line spectrum, the short wavelengths of light (or blue end) of the electromagnetic spectrum are shown on the left side and the long wavelengths of light (or red end) of the spectrum are shown on the right side.

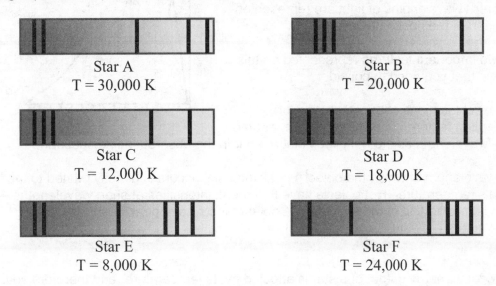

Star A
T = 30,000 K

Star B
T = 20,000 K

Star C
T = 12,000 K

Star D
T = 18,000 K

Star E
T = 8,000 K

Star F
T = 24,000 K

1) Do cold stars always appear to have a different (greater or fewer) number of lines in their absorption spectra than hot stars? Cite evidence from the above spectra to support your answer.

2) Do cold stars always appear to have more lines at either the blue or red ends of their absorption spectra than hot stars? Cite evidence from the above spectra to support your answer.

3) Consider the absorption line spectrum given below for Star G. Can you determine the approximate temperature for Star G by comparing its absorption line spectrum to the absorption line spectra and temperatures of Stars A – F given above? If so, write in your estimate in the space below; if not, explain why not.

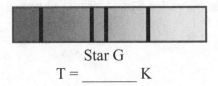

Star G
T = _____ K

The spectral curve on the graph at right illustrates the energy output versus wavelength for Star G. Again, the short (or blue) wavelengths of light are represented on the left side of the horizontal axis and the long (or red) wavelengths of light are represented on the right end of the horizontal axis.

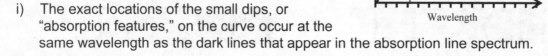
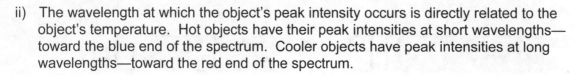

There are two important features represented on this spectral curve that you need to consider.

i) The exact locations of the small dips, or "absorption features," on the curve occur at the same wavelength as the dark lines that appear in the absorption line spectrum.

ii) The wavelength at which the object's peak intensity occurs is directly related to the object's temperature. Hot objects have their peak intensities at short wavelengths—toward the blue end of the spectrum. Cooler objects have peak intensities at long wavelengths—toward the red end of the spectrum.

Although the total energy output of a star is affected by its temperature (and therefore so is the height of the spectral curve), for this activity, we will assume that the height (but not location) of the peak intensity and the general shape of the spectral curves for Stars A-F can be drawn nearly the same for each star. Only the location of the peak intensity and the location of the small dips, or absorption features, will be different for each star.

4) Examine the spectral curve shown at right for Star A (the same Star A as on the previous page). Note that Star G has a temperature of approximately 25,000 K, while Star A has a temperature of 30,000 K. With this information, would you say the wavelength for the peak intensity of Star A is drawn at approximately the correct wavelength as compared to the wavelength for the peak of Star G? Explain your reasoning.

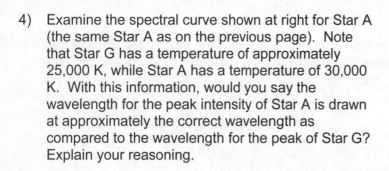
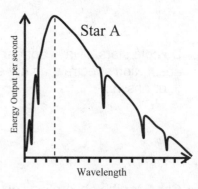

5) Are the absorption features (dips) in the spectral curve for Star A drawn at approximately the correct wavelengths? Explain how you can tell.

6) Sketch spectral curves for Stars B – F on the corresponding graphs provided below. Make sure your sketch includes the absorption features and the peak intensities drawn at *approximately* the correct wavelengths.

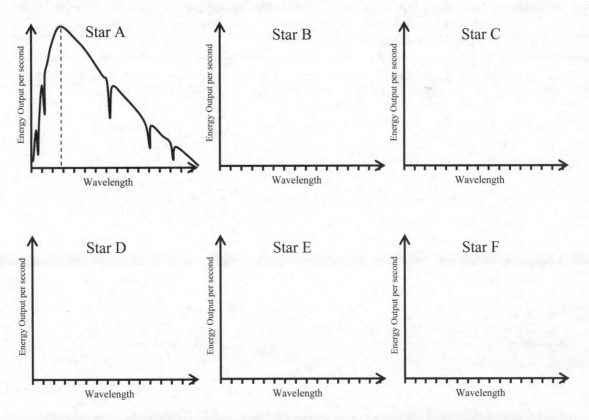

7) Did you draw the peak intensity of each spectral curve at the same wavelength as the spectral curve for Star A? Why or why not?

8) If you were given a star's absorption line spectrum and its corresponding spectral curve shown on an energy output per second versus wavelength graph, how could you approximate the temperature of the star?

9) Consider the following statement made by a student regarding a star's temperature and its corresponding absorption line spectrum.

Student: *If I am looking at a star's absorption line spectrum and see that it has a lot of lines at the blue end of the spectrum, then the star must be hot because the blue lines are higher energy lines.*

Do you agree or disagree with this student? Explain your reasoning and support your answer by citing evidence from the absorption line spectra given for Stars A – G.

The drawing below illustrates the amount that different wavelengths of light are able to penetrate Earth's atmosphere. The dotted/shaded regions are used in this drawing to depict the different layers of air in Earth's atmosphere. Notice that the atmosphere can be transparent to light at some wavelengths (all three lines passing through the atmosphere to the surface of Earth) and yet can also completely block other wavelengths of light (all three lines being absorbed in the atmosphere).

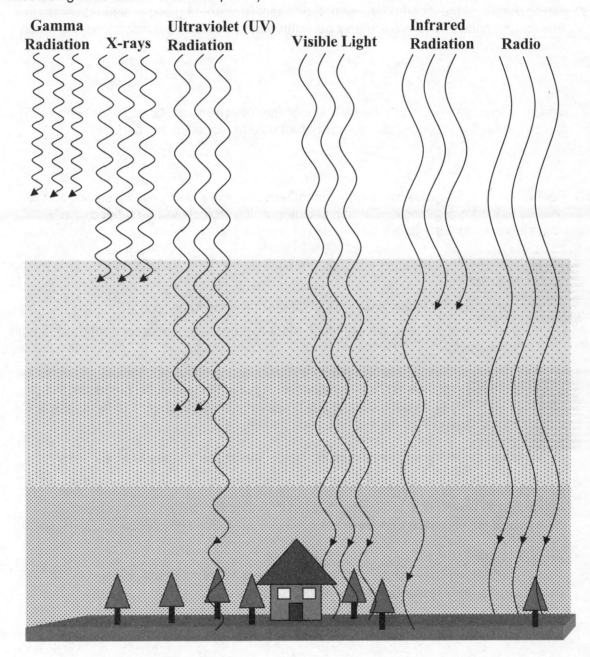

1) Which, if any, of the different wavelengths of light (electromagnetic radiation) shown in the image on the previous page are able to **entirely** penetrate Earth's atmosphere and reach the surface?

2) Which, if any, of the different wavelengths of light (electromagnetic radiation) shown in the image on the previous page only **partially** penetrate Earth's atmosphere and reach the surface?

3) Which, if any, of the different wavelengths of light (electromagnetic radiation) shown in the image on the previous page are **not at all** able to reach Earth's surface?

4) Federal funding agencies must form committees to decide which telescope projects will receive funds for construction. When deciding which projects will be funded, the committees must consider:

   i.   that certain wavelengths of light are blocked from reaching Earth's surface by the atmosphere,

   ii. how efficiently telescopes work at different wavelengths, and

   iii. that telescopes in space are much more expensive to construct than Earth-based telescopes.

   Use these three criteria when you consider each pairing of telescope proposals listed below (a – d). For **each pair,** state which proposal you would choose to fund. Explain the reasoning behind your decision for each pair.

   a) Which of the two proposals described below would you choose to fund? Why?

   Project Delta:
   A gamma ray wavelength telescope, located in Antarctica, will be used to look for evidence of the presence of a black hole.

   Project Theta:
   A visible wavelength telescope, located on a university campus, will be used in the search for planets outside the solar system.

b) Which of the two proposals described below would you choose to fund? Why?

Project Beta:
An X-ray wavelength telescope, located near the North Pole, will be used to examine the Sun.

Project Alpha:
An infrared wavelength telescope, placed on a satellite in orbit around Earth, will be used to view supernovae.

c) Which of the two proposals described below would you choose to fund? Why?

Project Rho:
A UV wavelength telescope, placed high atop Mauna Kea in Hawaii at 14,000 ft above sea level, will be used to look at distant galaxies.

Project Sigma:
A visible wavelength telescope, placed on a satellite in orbit around Earth, will be used to observe a pair of binary stars located in the constellation Ursa Major.

d) Which of the two proposals described below would you choose to fund? Why?

Project Zeta:
A radio wavelength telescope, placed on the floor of the Mojave desert, will be used to detect potential communications from distant civilizations outside our solar system.

Project Epsilon:
An infrared wavelength telescope, located in the high elevation mountains of Chile, will be used to view newly forming stars (protostars) in the Orion nebula.

The most difficult part of constructing an accurate model for planetary motions is that planets seem to wander among the stars. During their normal (or prograde) motion, planets appear to move from west to east over many consecutive nights as seen against the background stars. However, they occasionally (and predictably) appear to reverse direction and move east to west over consecutive nights as seen against the background stars. This backwards motion is called retrograde motion.

1) Given the data in Table 1, plot the motion of the mystery planet on the graph provided in Figure 1 (record dates next to each position you plot). Then, draw a smooth line, using your data points, to illustrate the path of the planet through the sky.

Date of Observation	Azimuth (horizontal direction)	Altitude (vertical direction)
May 1	240	45
May 15	210	50
June 1	170	50
June 15	150	45
July 1	170	40
July 15	180	45
August 1	140	50
August 15	120	55

Table 1 – Mystery Planet Positions

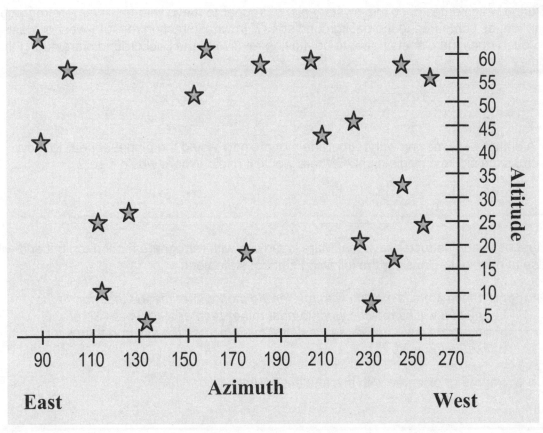

Figure 1 – Planet Path

For each date, time of observation changed so the stars would keep same Az/Alt

2) On what date was the mystery planet located farthest to the west? What was the azimuth of the planet on this date?

3) On what date was the mystery planet located farthest to the east? What was the azimuth of the planet on this date?

4) Describe how the mystery planet moved (east or west), as compared to the background stars, during the time between the dates identified in questions 2 and 3.

5) During which dates does the mystery planet appear to move with normal, prograde, motion, as compared to the background stars? In what direction (east-to-west or west-to-east) does the planet appear to be moving relative to the background stars during this time?

6) During which dates does this mystery planet appear to move with backward, retrograde, motion, as compared to the background stars? In what direction (east-to-west or west-to-east) does the planet appear to be moving relative to the background stars during this time?

7) If a planet were moving with retrograde motion, how would the planet appear to move across the sky in a single night? Where would it rise? Where would it set?

8) Suppose your instructor says that Mars is moving with retrograde motion tonight and will rise at midnight. Consider the following student statement:

Student: *Since Mars is moving with retrograde motion that means that during the night it will be moving west-to-east rather than east-to-west. So at midnight it will rise in the west and move across the sky and then later set in the east.*

Do you agree or disagree with this student? Why?

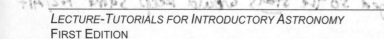

In this activity we will investigate the relationship between how long it takes a planet to orbit a star (orbital period) and how far away that planet is from the star (orbital distance). We will start by investigating an imaginary planetary system that has an average star like the Sun at the center. A huge Jupiter-like, Jovian planet named Moto orbits close to the star, while a small Earth-like, terrestrial planet named Spec is in a far away orbit around the star. Use this information when answering the next four questions. If you're not sure of the correct answers to questions 1 – 4, just take a guess. We'll return to these questions later in this activity.

1) Which of the two planets (Moto or Spec) do you think will move around the central star in the least amount of time? Why?

2) If Moto and Spec were to switch positions, would your answer to question 1 change? If so, how? If not, why not?

3) Do you think the orbital period for Moto would increase, decrease, or stay the same if it were to move from being close to the central star to being much farther away? Why?

4) Imagine both Moto and Spec were in orbit around the central star at the same distance and that their orbital positions would never intersect (so that they would never collide). Do you think the two planets would have the same or different orbital periods? Why?

The graph below illustrates how the orbital period (expressed in years) and orbital distance (expressed in Astronomical Units, AU) of a planet are related. *orbiting our sun*

5) According to the graph, would you say that a planet's orbital period appears to increase, decrease, or stay the same as a planet's orbital distance is increased?

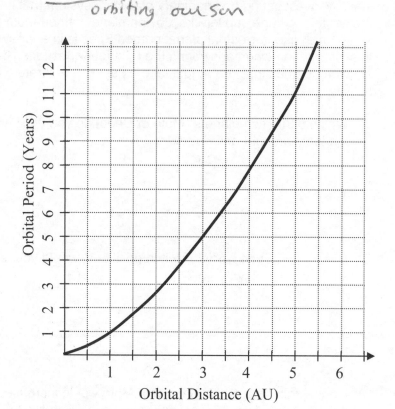

6) How far from the central star does a planet orbit if it has an orbital period of 1 year?

7) How long does it take a planet to complete one orbit if it is twice the distance from the central star as the planet described in question 6?

8) Based on your results from questions 6 and 7, which of the following best describes how a planet's orbital period will change (if at all) when its distance to the central star is doubled? Circle your choice.

    a) The planet's orbital period will decrease by half.

    b) The planet's orbital period will not change.

    c) The planet's orbital period will double.

    d) The planet's orbital period will more than double.

In the table below we have provided the orbital distances, orbital periods and masses for the six planets closest to the Sun.

Planet	Orbital Distance (in Astronomical Units - AU)	Orbital Period (in Years)	Planet mass (in units of Earth's mass)
Mercury	0.38	0.24	0.06
Venus	0.72	0.61	0.82
Earth	1.0	1.0	1.0
Mars	1.52	1.88	0.11
Jupiter	5.20	11.86	318
Saturn	9.54	29.46	95.2

9) What is the name of the planet that you identified the orbital distance for in question 6?

10) Consider the information provided in the table and on the graph and choose the answer below that best describes the effect that a planet's mass has on its orbital period. Circle your choice.

a) Planets that have small masses have longer orbital periods than planets with large masses.

b) Planets with the same mass will also have the same orbital period.

c) Planets that have large masses have longer orbital periods than planets with small masses.

d) A planet's mass does not affect the orbital period of a planet.

Explain your reasoning and cite a specific example from the table or graph to support your choice.

11) Review your answers to questions 1 – 4. Do you still agree with the answers you provided? If not, describe (next to your original answers) how you would change the answers you gave initially.

Debrief: Planet Mass would matter for a planet with

huge mass - like 10 Jupiter masses.

The extremely high temperature of Earth's core causes material in the surrounding mantle to become hot, expand, and rise toward the surface. The mantle material then cools and sinks, resulting in a circular motion of material moving beneath Earth's surface. This circulation of mantle material causes the continental and oceanic plates to move across Earth's surface. At various locations on Earth's surface, we are able to observe plates colliding, plates separating, and plates moving horizontally.

The drawing below shows a cross section of Earth's surface and its underlying mantle. At this particular location of the surface, the dense oceanic plate is being forced beneath the less dense continental crust. The dense oceanic plate experiences higher temperatures (and pressures) as it is forced deeper into the mantle. This interaction between the oceanic plate and continental plate causes molten material to move upward through the continental plate until it breaks the surface in the form of volcanoes.

Oceanic to Continental Plate Convergence Zone

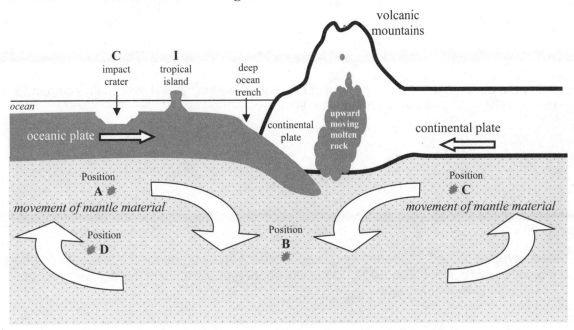

1) In the drawing above, which way (*right or left*) are the oceanic and continental plates moving?

2) Which is hotter, the piece of mantle material at Position A or the piece of mantle material at Position D? Explain your reasoning.

3) What direction are the pieces of mantle material moving (up, down, left, or right) at Positions A, B, C, and D?

---

4) Consider the following statements made by two students debating why the oceanic and continental plates move.

   Student 1: *The plates are moving because the mantle material is constantly moving beneath Earth's plates, and this causes the plates to move.*

   Student 2: *I disagree. The plates are just floating on the mantle material. The plates started moving a long time ago when Earth initially formed and the plates' momentum keeps them moving toward each other.*

   Do you agree or disagree with either or both of the students? Why?

5) Just beneath *point I* on the drawing is a tropical island. What will eventually happen to the island as the oceanic plate moves? Why?

6) Just beneath *point C* on the drawing is an ancient impact crater on the ocean floor where a giant comet collided with Earth. What will happen to the ancient impact crater as the oceanic plate moves? Why?

7) Imagine that an impact occurred on the continental plate millions and millions of years ago, leaving behind an impact crater near the right side of the base of the volcano. Why would there be little evidence of this impact crater found today?

8) Consider the image below of the rocky and crater-covered Moon. Its very old surface has remained virtually unchanged over the last few billion years. Do you think the Moon has an active, hot and molten interior or an inactive, cold and solid interior? Why?

9) If a new planet were discovered, what evidence would you look for to determine whether or not it has an active, hot and molten interior? Why?

Consider the information provided in the graph and table below. The graph shows the temperature (expressed in Kelvin) at different distances from the Sun (expressed in Astronomical Units, AU) in the solar system during the time when the planets were originally forming. The table provides some common temperatures to use for comparison.

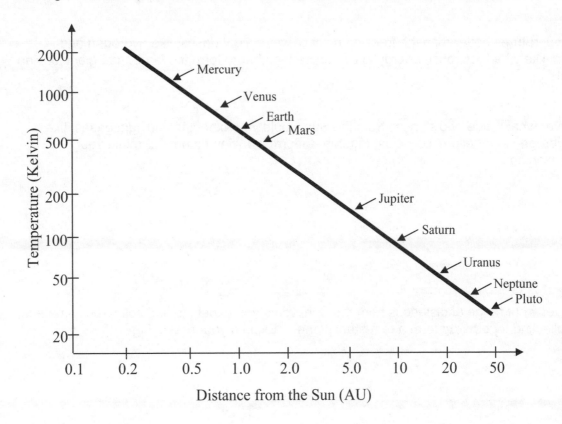

Condition	Temp. Fahrenheit	Temp. Celsius	Temp. Kelvin
Severe Earth Cold	-100	- 73	199
Water Freezes	32	0	273
Room Temp	72	22	296
Human Body	98.6	37	310
Water Boils	212	100	373

1) What was the temperature at the location of Earth?

2) What was the temperature at the location of Mars?

3) Which planets formed at temperatures hotter than the boiling point of water?

4) Which planets formed at temperatures cooler than the freezing point of water?

At temperatures hotter than the freezing point of water, light gases, like hydrogen and helium, likely had too much energy to condense together to form the large, gas-giant, Jovian planets.

5) Over what range of distances from the Sun would you expect to find light gases, like hydrogen and helium, collecting together to form a Jovian planet? Explain your reasoning.

6) Over what range of distances from the Sun would you expect to find solid, rocky material collecting together to form a terrestrial planet? Explain your reasoning.

7) Is it likely that a large, Jovian planet would have formed at the location of Mercury? Explain your reasoning.

Part I: Earth and Moon

Astronomers often deal with large numbers for distances, masses, and other quantities. They often use ratios to get a better sense of how big or small these quantities are. This can be useful in our daily lives as well. For example, we may not have a good sense for the length of a 40-meter-long commercial jet, but saying that the jet is 8 times longer than a car may be more meaningful to us. In this activity we will use ratios to try to better understand the size of objects in the solar system, in particular the size of the Sun.

Distances like the following can be hard to conceptualize:

Moon's diameter: 3,476 km
Earth's diameter: 12,756 km

But we can think about these sizes in terms of one another so that we can create a scale model of Earth and the Moon in our minds. If we wish to express how many times bigger Earth is than the Moon, we can divide Earth's diameter by the Moon's diameter. The result is roughly 4 (12,756/3476 ≈ 4). This means Earth is approximately four times wider than the Moon, or equivalently, you could fit about four Moons across the diameter of Earth (as shown below).

1) Which of the following pairs of objects would make a good scale model of Earth and the Moon?

    a) a basketball and a soccer ball
    b) a basketball and a baseball (or softball)
    c) a basketball and a ping pong ball
    d) a basketball and a pea
    e) a basketball and a grain of sand

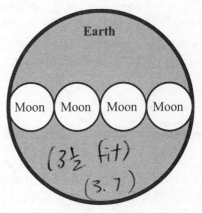

The distance between Earth and the Moon is much larger than either the Moon or Earth – but how much larger? If we divide the distance between Earth and the Moon (384,000 km) by Earth's diameter, we get 384,400/12,756 ≈ 30. This means you could fit approximately 30 Earths in the space between Earth and the Moon.

2) Sketch a scale model of the Earth-Moon orbital system below. When making your sketch, use the two scale ratios described above.

3) To make a scale model of the Earth-Moon orbital system, you not only need to pick appropriately sized objects to represent Earth and the Moon; you also need to place them the correctly scaled distance apart. Let's say you use a 1-foot (12-inch) basketball and a 3-inch orange as your Earth and Moon respectively. About how far apart must they be placed to represent an accurate scale model of the Earth-Moon orbital system? (Circle your answer below.) Why?

   a) 1 foot
   b) 4 feet
   c) 10 feet
   d) 30 feet
   e) 300 feet

Part II: The Sun

no diagram?

Compared to the size of Earth, the Sun (with a diameter 1,392,000 km) is about 110 times bigger than Earth (1,392,000/12,756 ≈ 110).

4) Can any combinations of the following items be used to make an accurate scale model of Earth and the Sun? If so, which two would you choose and why? If not, why not?
   - basketball
   - soccer ball
   - baseball (or softball)

   - ping pong ball ✗
   - pea
   - grain of sand

Now let's compare the Sun's diameter to the size of the Moon's orbit around Earth. The diameter of the Moon's orbit around Earth is about 769,000 km across. So, the ratio of the Sun's diameter to the Moon's orbital diameter is roughly 2. (1,392,000/769,000 ≈ 2)

5) Does this mean that two Suns placed side by side would fit inside the Moon's orbit around Earth, or that two Moon orbits placed side-by-side would fit across the Sun? Draw a sketch below to illustrate your answer.

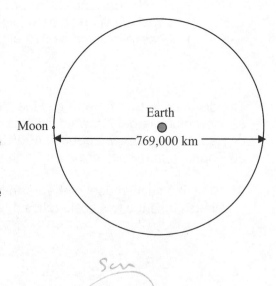

Moon

Earth

769,000 km

The distance from Earth to the Sun is about 150,000,000 km. This makes the distance between the Sun and Earth about 110 times larger than the diameter of the Sun (150,000,000/1,392,000 ≈ 110).

6) If you were to use a 1-foot (12-inch) basketball to represent the Sun, how far would it have to be from Earth to be an accurate scale model?
   a) 1 foot
   b) 10 feet
   c) 30 feet
   d) 110 feet
   e) 300 feet

7) If we used a basketball to represent the Sun and a ping-pong ball to represent Earth, and separated them by the distance you answered in question 6, would we have an accurate scale model of the Sun-Earth system? Explain your answer.

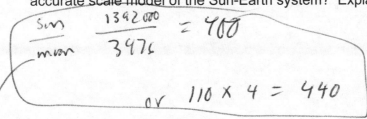

Sun/moon 1392000/3476 = 400

or 110 x 4 = 440

No, ⊕ not that big

8) How many Moons would fit across the diameter of the Sun?

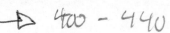

→ 400 - 440

9) Approximately how many times could the Moon's orbital diameter fit between Earth and the Sun?

110 suns fit (top of page)

← 2 moon diameters across sun

220

110 Earths across Sun

110 Suns from here to Earth

$$10\,pc \qquad m = M$$

$$> 10\,pc \qquad m \text{ dimmer than } M$$
$$m > M \text{ numerically}$$

(First teach what they are)
Emphase   skip this page —
Apparent and Absolute Magnitudes of Stars   (13 min) 67
                                                          this page

1) Which value, apparent magnitude or absolute magnitude, do you think:

   a) tells us how bright an object will appear from Earth?

   Ap

   b) tells us about the object's actual brightness?

   Ab

2) Consider the following debate between two students.

   * **Student 1:** *I think a star with an apparent magnitude of –2.0 would look brighter than a star with an apparent magnitude of +1.0.*
   Student 2: *I disagree. You don't understand the number scale for apparent and absolute magnitude. The bigger the number the brighter the star. So the +1.0 star would look brighter than the –2.0 star.*

   Do you agree or disagree with either or both students? Why?

3) Star Y appears much brighter than Star Z when viewed from Earth, but is found to actually give off much less light. Assign a set of possible values for the apparent and absolute magnitudes of these stars that would be consistent with the information given in the previous statement.

	App	Abs
Y	1	10
Z	5	8

4) The star Rigel has an apparent magnitude of 0.1 and is located about 250 parsecs away from Earth. Which of the following is most likely the absolute magnitude for Rigel?

   a) -6.9
   b) 0.1
   c) 7.1

   Explain your reasoning.

   Very hard —
   explain that farther
   away then 10pc

   $m > M$ *numerically*

   Covered better on p. 79-8

5) Refer to the following table for questions 5(a)-5(d):

	Apparent Magnitude	Absolute Magnitude
Star A:	1	1
Star B:	1	2
Star C:	5	4
Star D:	4	4

a) Which object appears brighter from Earth: Star C or Star D? Explain your reasoning. D

b) Which object is actually brighter: Star A or Star D? Explain your reasoning.

Same

c) For Stars A – D state whether the star is closer than, farther than, or exactly 10 parsecs away from Earth. Explain your reasoning.

A - 10 pc
B - farther
C - closer.
D - 10 pc

d) Would the apparent magnitude of Star A increase, decrease, or stay the same, if it were located at a distance of 40 parsecs? What about the absolute magnitude? Explain your reasoning.   \Same

dimmer -
smaller m

6) Star F is known to have an apparent magnitude of –26.7 and an absolute magnitude of 4.8. Where might this star be located? What is the name of this star? Explain your reasoning.

Part I: Stars in the Sky

Consider the diagram to the right.

Distant stars

1) Imagine that you are looking at the stars from Earth in January. Use a straightedge or a ruler to draw a straight line from Earth in January, through the nearby Star A, out to the distant stars. Which of the distant stars would **appear** closest to the nearby Star A in your night sky in January? Circle this star and label it "Jan."

☆
Nearby star
(Star A)

2) Repeat question 1 for July and label the star "July."

3) Below, these same distant stars are shown as you would see them in the night sky. In this box, draw a small × to indicate the position of Star A as seen in January and label it "Jan."

4) In the same box, draw another × to indicate the position of Star A as seen in July and label it "July."

5) Describe the motion of Star A relative to the distant stars as Earth orbits the Sun counterclockwise from January of one year, through July, to January of the following year.

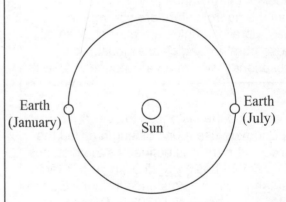

Earth
(January)

Sun

Earth
(July)

arcsecods.

The Parsec

The apparent motion of nearby objects relative to distant objects, which you just described, is called **parallax**.

6) Consider two stars (C and D) that both exhibit parallax. If Star C appears to move back and forth by a greater amount than Star D, which star is closer to you? If you're not sure, just take a guess. We'll return to this question later in this activity.

Part II: What's a Parsec?

Consider the diagram to the right.

7) Starting from Earth in January, draw a line through Star A to the top of the page. use p71 as straight edge

8) There is now a narrow triangle with the Earth-Sun distance as its base. The small angle, just below Star A, formed by the two longest sides of this triangle is called the **parallax angle** for Star A. Label this angle "p_A."

Knowing a star's parallax angle allows us to calculate the distance to the star. Since even the nearest stars are still very far away, parallax angles are extremely small. These parallax angles are measured in "arcseconds" where an arcsecond is 1/3600 of 1 degree.

To describe the distances to stars, astronomers use a unit of length called the **parsec**. 1 parsec is defined as the distance to a star that has a **par**allax angle of exactly 1 arc**sec**ond. The distance from the Sun to a star 1 parsec away is 206,265 times the Earth-Sun distance or 206,265 AU. (Note that the diagram to the right is not drawn to scale.)

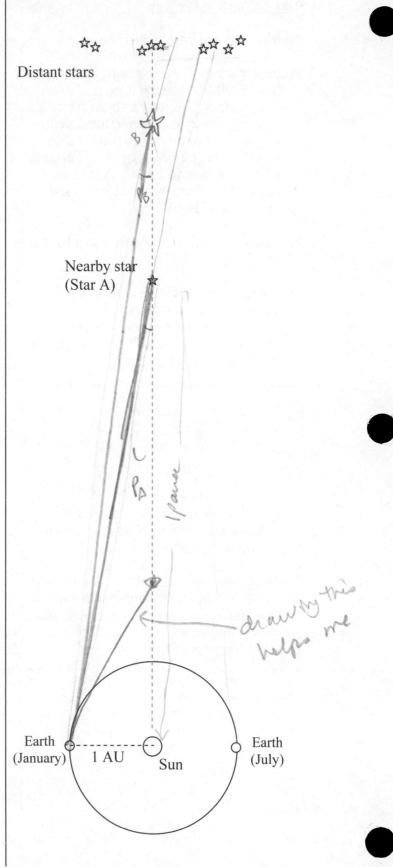

Distant stars

Nearby star
(Star A)

draw by this helps me

Earth
(January)   1 AU   Sun   Earth
(July)

9) If the parallax angle (for Star A) p_A is 1 arcsecond, what is the distance from the Sun to Star A? (Hint: use parsec as your unit of distance.) Label this distance on the diagram.

10) Is a parsec a unit of length or a unit of angle? It can't be both.

Note: Since the distance from the Sun to even the closest star is so much greater than 1 AU, we can consider the distance from Earth to a star and the distance from the Sun to that star to be approximately equal.

Part III: Distances

11) Consider the following discussion between two students working on this tutorial regarding the relationship between parallax angle and the distance we measure to a star:

Student 1:   *If the distance to the star is more than 1 parsec, then the parallax angle must be more than 1 arcsecond. Larger distance means larger angle.*

Student 2:   *If we drew a diagram for a star that was much more than 1 parsec away from us, the triangle in the diagram would be pointier than the one we just drew in Part II. That should make the parallax angle smaller for a star farther away.*

Do you agree or disagree with either or both of the students? Why?

12) On your diagram from Part II, draw a second star on the dotted line farther from the Sun than Star A. Do this in another color, if possible, and label it "Star B." Repeat steps 7 and 8 from Part II, except label the parallax angle p_B.

13) Which star, the closer one or the farther one, has the larger parallax angle? Check your answer against your answers for questions 6 and 11 and resolve any discrepancies.

✱ Be sure they see The farther ✱
has a smaller angle

$$(\text{give } l\text{-}t\text{-}h\text{-})$$
$$p = \frac{1}{d}$$

Part I: Angular Measurement

Imagine that you are standing in an open field. While facing south, you see a house in the distance. If you look to the east, you see a barn in the distance.

1) What is the angle between the house and the barn? (Hint: If you point at the barn with one arm and point at the house with your other arm, what angle do your arms make?)

2) You see the Moon on the horizon just above the barn in the east, and also see a bright star directly overhead. What is the angle between the Moon and the overhead star?

3) Compare your answers for the barn-house angle from question 1 and the Moon-star angle from question 2. Are they the same? Does this separation angle tell you anything about the actual distance between the barn and house or the Moon and star?

We are unable to **directly** measure distances to objects in our night sky. However, we can obtain the distances to relatively nearby stars by using their parallax angles. Because even these stars are very far away (up to about 500 parsecs), the parallax angles for these stars are very small. They are measured in units of **arcseconds**, where 1 arcsecond is 1/3600 of one degree. To give you a sense of how small this angle is, the thin edge of a credit card, when viewed from one football field away, covers an angle of about 1 arcsecond.

Part II: Finding Stellar Distance Using Parallax

Consider the star field drawing shown in Figure 1. This represents a tiny patch of our night sky. In this drawing we will imagine that the angle separating Stars A and B is just ½ of an arcsecond.

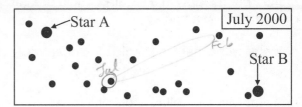

Figure 1

In Figure 2 (see the final page of the activity) there are pictures of this star field taken at different times during the year. One star in the field exhibits parallax as it moves back and forth across the star field with respect to the other, more distant stars.

4) Using Figure 2, determine which star exhibits parallax. Circle that star on each picture in Figure 2.

5) In Figure 1, draw a line that shows the range of motion for the star exhibiting parallax in the pictures from Figure 2. Label the endpoints of this line with the months when the star appears at those endpoints.

6) How many times bigger is the separation between Stars A and B compared to the distance between the endpoints of the line showing the range of the motion for the star exhibiting parallax?   2

7) Recall that Stars A and B have an angular separation of ½ of an arcsecond in Figure 1. Consider two more stars (C and D) that are separated **twice** as much as Stars A and B. What is the angular separation between Stars C and D in arcseconds?

      1 "

8) What is the angular separation between the endpoints that you marked in Figure 1 for the nearby star exhibiting parallax?   ½" $\frac{1}{4}$ "

Note: We define a star's parallax angle as *half* the angular separation between the endpoints of the star's angular motion.

9) What is the parallax angle for the nearby star from question 8?

      $\frac{1}{8}$"

Note: We define 1 **parsec** as the distance to an object that has a **par**allax angle of 1 arc**sec**ond. For a star with a parallax angle of 2 arcseconds, the distance to the star from Earth would be ½ of a parsec.

10) For a star with a parallax angle of ½ of an arcsecond, what is its distance from us?

      2 parsec

11) For a star with a parallax angle of ¼ of an arcsecond, what is its distance from us?

      4 parsec

12) What is the distance from us to the nearby star exhibiting parallax in the pictures from Figure 2? (Hint: consider your answer to question 9.)
    a) 1 parsec
    b) 2 parsecs
    c) 4 parsecs
    d) 8 parsecs
    e) 16 parsecs

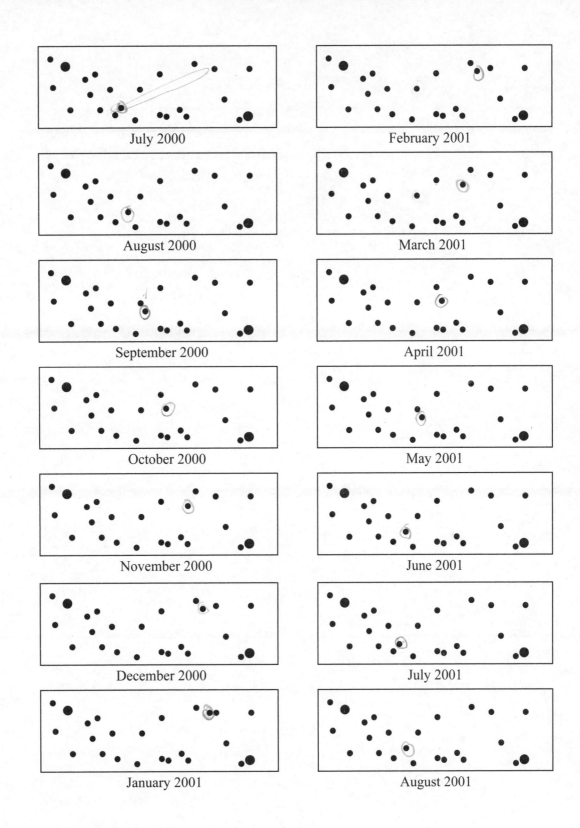

July 2000

August 2000

September 2000

October 2000

November 2000

December 2000

January 2001

February 2001

March 2001

April 2001

May 2001

June 2001

July 2001

August 2001

Figure 2

The H-R diagram below will be used to answer questions throughout this activity.

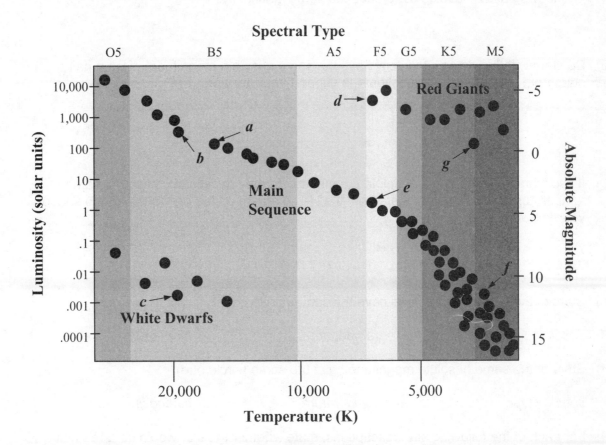

1) What are the spectral type, temperature, absolute magnitude and luminosity of Star **a**?

Spectral type:

Temperature:

Absolute magnitude:

Luminosity:

2) Which two pairs of labeled stars in the diagram have the same temperature?

3) Do stars of the same temperature have the same spectral type? Use a pair of stars from your answer to question 2 to support your answer.

4) Which two pairs of labeled stars have the same luminosity?

5) Do stars with the same luminosity have the same absolute magnitude? Use a pair of stars from your answer to question 4 to support your answer.

6) If two stars have the same absolute magnitude, do they necessarily have the same temperature? Use a pair of stars from the H-R diagram on the previous page to support your answer.

7) Stars of the same spectral type have the same (circle one):

   absolute magnitude   *temperature*   *luminosity*

8) Stars of the same absolute magnitude have the same (circle one):

   spectral type   *temperature*   *luminosity*

9) For each of the following star descriptions, state whether the star would be a red giant, white dwarf, or main sequence star, and provide the letter(s) of a star from the H-R diagram that fits each description.

   a) very bright (high luminosity) and very hot (high temperature)

   b) very dim and cool

   c) very dim and very hot

   d) very bright and cool

~ Swin this page, 10 ~ 15 min total

Part I: Magnitudes and Star Distances

Below is a table of four stars and their apparent and absolute magnitudes. Use this table to answer the following questions.

	Apparent Magnitude	Absolute Magnitude	Distance
Star A:	0	0	10 pc
Star B:	0	2	< 10
Star C:	5	4	> 10 pc
Star D:	4	4	10

1) Which object appears brighter from Earth: Star C, Star D, or neither? Explain your reasoning.

   D

2) Which object is more luminous: Star C, Star D, or neither? Explain your reasoning.

   Same — Abs mag.

3) Star B has an apparent magnitude of 0, which tells us how bright it appears from Earth at its true location. Star B has an absolute magnitude of 2, which tells us how bright it would appear if it were at a distance of 10 parsecs (about 33 light-years).

   Where would Star B appear brighter, in its true location or if it were at a distance of 10 parsecs? Explain your reasoning. brighter: 0 mag — true location

4) Is Star B *closer than 10 pc*, *farther than 10 pc* or *exactly 10 pc* away? Record your answer in the table above and explain your reasoning.

   Closer since dimmer at 10pc

5) How far away is Star D? Record your answer in the table above.

   10 pc since abs = app

6) Is Star C *closer than 10 pc*, *farther than 10 pc* or *exactly 10 pc* away? Record your answer in the table above.

7) Complete the remaining blank in the distance column of the above table and check your answers with another group.

LECTURE-TUTORIALS FOR INTRODUCTORY ASTRONOMY
FIRST EDITION

Tell them its |A0 — A9|
← cooler

Part II: Spectroscopic Parallax

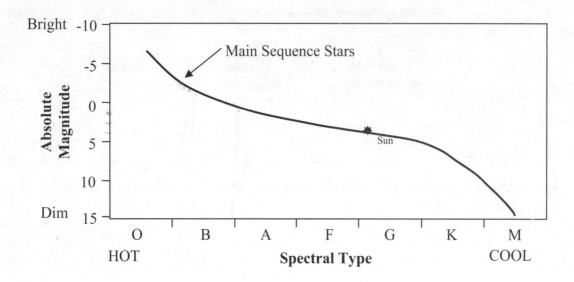

Below is a table giving both the apparent magnitude and spectral type for five **main sequence** stars. For each star, do the following:

8) Using the above H-R diagram, estimate the absolute magnitude for each star and write your answer in the absolute magnitude column of the table below.

9) Complete the distance column in the table below by classifying each star as being **closer**, **slightly farther**, or **much farther** than 10 parsecs away. This procedure, called spectroscopic parallax, provides astronomers with another strategy to measure the distance to stars.

Star	Apparent Magnitude	Spectral Type	Absolute Magnitude	Distance Estimate	
Rigel Kentaurus	0.0	G2	3 – 4	closer	<10
Vega	0.04	A0	1 or 0	~10 pc	
Rigel B	6.6	B9	0	Much farther	
Achernar	0.5	B3	–2	farther	
Tau Scorpius	2.8	B0	–4	much farther	

Note: *By completing this table, you have estimated the distance to a star by comparing the apparent and absolute magnitudes without using a formula. The exact distance can be calculated using the formula $d=10^{(m-M+5)/5}$ pc, where m is the apparent magnitude and M is the absolute magnitude. It is not necessary to perform any calculations to complete the table above.*

contracting gas heats up

~8 min

Stars begin life as a cloud of gas and dust. The birth of a star begins when a disturbance, such as the shockwave from a supernova, triggers the cloud of gas and dust to spin and collapse inward.

1) Imagine that you are observing the region of space where a cloud of gas and dust is beginning to collapse inward to form a star (the object that initially forms in this process is called a protostar). Will the atoms in the collapsing cloud move away from one another, move closer to one another or stay at the same locations?

2) What physical interaction or force causes the behavior described in question 1 to occur?

gravity

3) Would you expect the temperature of the protostar to increase or decrease with time? Explain your reasoning.

✳

↓

The inward collapse of material causes the center of the protostar to become very hot and dense. Once the central temperature and density reach critical levels, nuclear fusion reactions begin. During the fusion reaction, hydrogen atoms are combined together to form helium atoms. When this happens, photons of light are emitted. Once these reactions are sustained, the protostar becomes a main sequence star. At this time the star no longer collapses, and a state of *hydrostatic equilibrium* is reached between the inward gravitational collapse of material and the outward pressure caused by the energy given off during nuclear fusion reactions.

Consider the information shown in the table below when answering questions 4 through 7.

1000 million = 1 billion

Mass of the Star (in multiples of Sun masses, M_{Sun})	Main Sequence Lifetime of the Star
0.67 M_{Sun}	45 billion years
1.0 M_{Sun}	10 billion years —
1.3 M_{Sun}	800 million years
2.0 M_{Sun}	500 million years
6.0 M_{Sun}	70 million years
60 M_{Sun}	800 thousand years

4) Which live longer, high-mass or low-mass stars?

5) Based on your answer to question 4, do you think that the nuclear fusion rate for a high-mass star is greater than, less than, or equal to the nuclear fusion rate of a low-mass star?

6) Which of the following statements best describes how the lifetimes compare between Star A (a star with a mass equal to the Sun) and Star B (a star with six times the mass of the Sun)? Circle the best possible response given below. (Note: it may be helpful to examine the information given in the table on the previous page.)

   a) Star A will live less than 1/6th as long as Star B.
   b) Star A will live 1/6th as long as Star B.
   c) Star A will have the same lifetime as Star B.
   d) Star A will live six times longer than Star B.
   e) Star A will live more than six times longer than Star B.

   Explain your reasoning for the choice you made.

7) The Sun has a lifetime of approximately 10 billion years. If you could determine the rate of nuclear fusion for a star with twice the mass of the Sun, which of the following would best describe how its fusion rate would compare to the Sun? Circle the best possible response to complete the sentence given below. (Note: it may be helpful to examine the information given in the table on the previous page.)

   A star with twice the mass of the Sun would have a nuclear fusion rate that is _____ the fusion rate of the Sun.

   a) less than
   b) a little more than
   c) twice
   d) more than twice

Explain your reasoning for the choice you made.

Main sequence stars that can no longer support nuclear fusion of hydrogen in their cores will become red giant stars. Although most main sequence stars become red giants, their specific evolutionary paths after this red giant phase vary greatly depending on mass.

A low-mass star, less than about eight times the mass of our Sun, eventually ejects its outer layers to produce a planetary nebula. The stellar core remaining in the middle of this planetary nebula is called a white dwarf. If the white dwarf is isolated, it will very slowly cool until it becomes a black dwarf.

In contrast, a high-mass star, more than eight times the mass of the Sun, will eventually explode as a Type II supernova. Depending on the original mass of the star, the Type II supernova will leave behind either a neutron star or, if the original star was extremely massive, a black hole.

1) Use the information above and the word list below to fill in the ovals in the diagram on the next page. Be sure to look at the arrows and words between the ovals to make sure these links between ovals make sense. Check your work with another group.

> Word list:
> neutron star
> black hole
> planetary nebula
> white dwarf
> nova
> Type II supernova

The diagram does not give us all the information known about the death of stars. Since it is incomplete, we can always add to this diagram as we learn more information.

2) In parts a and b below, you are given some additional information about the end-states of stars. It is your job to change or add to the diagram to incorporate this additional information. (Note: There are several ways to accomplish this.)

a) If a white dwarf has a nearby companion star, it can gravitationally attract material from its companion in a process known as accretion. When the white dwarf accretes enough material from the companion, the white dwarf will either explosively blow-off the outer layers as a nova, leaving behind the white dwarf unchanged, or violently explode as a Type I supernova, leaving nothing behind.

b) In isolated space, a black hole can be nearly impossible to detect. However, if a black hole has a binary companion star, the strong gravitational pull of the black hole can accrete matter from its companion. This material spirals around the black hole. This process causes the rapidly moving material to emit large amounts of X-ray radiation, which we can detect with X-ray telescopes. Thus, one way to look for black holes is to look for strong X-ray sources.

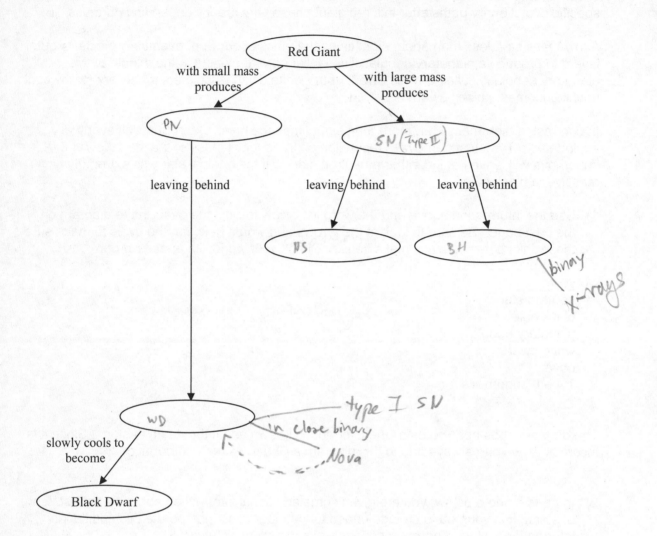

Red Giant

with small mass
produces

with large mass
produces

PN

SN (Type II)

leaving behind

leaving behind

leaving behind

NS

BH

binary
x-rays

WD

type I SN

in close binary

Nova

slowly cools to
become

Black Dwarf

This tutorial will give you a better understanding of the size of the Milky Way Galaxy by investigating the distances to objects within the Galaxy and to other objects in the Universe. Below is a picture of a spiral galaxy. This is a picture of NGC 3184, a spiral galaxy similar to the Milky Way. Because we are located within the Milky Way, we are unable to take a picture of our entire galaxy from the outside. Let's assume that this picture represents our Milky Way Galaxy and has the dimensions labeled below. **Note that in this picture, 1 centimeter (cm) represents 10,000 light-years (ly); equivalently you can use 1 millimeter (mm) to represent 1,000 light-years (ly).**

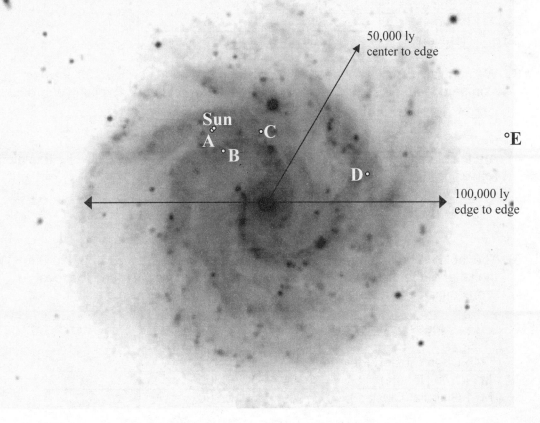

1) The Sun's position in the Milky Way is shown in the picture above. What is the approximate distance from the Sun to the center of the Milky Way? Recall that 1.0 cm represents 10,000 ly.

30,000 Ly

Sagittarius gal

LMC

2) The table below lists five bright stars in the night sky. Write the letter of the dot (A – E) from the picture above that best represents the location of each star. You can use letters more than once. Recall that 1 mm represents 1,000 ly.

If student are wrong here – don't discuss it with them until after #5

Star	Distance from Sun (in light years)	Letter
Sirius	9	A
Vega	26	A
Spica	260	A
Rigel	810	A
Deneb	1,400	A

3) We normally consider Deneb to be a bright but distant star at 1,400 ly away. Compared to the size of our galaxy, is Deneb truly distant? Explain your reasoning.

No, it is nearby if you compared to the whole galaxy.

4) Are the stars from question 2 inside or outside the Milky Way Galaxy? Explain your reasoning.

All near us – so in our galaxy.

5) The table below lists three Messier objects and their distances from the Sun. Write the letter of the dot (A – E) from the picture on the previous page that best represents the location of each object. You can use letters more than once.

Messier object	Distance from Sun (in light years)	Letter
M45 Open Cluster (Pleiades)	380	A
M1 (Crab Nebula)	6300	A
M71 Globular Cluster	12,700	B

6) Are these Messier objects part of the Milky Way? Explain your reasoning.

Yes, near us, so in our galaxy

7) The Crab Nebula has a width of about 11 light years. If you wanted to accurately draw the Crab Nebula on your diagram, would you use a small blob or a tiny dot at the location you indicated? Explain your reasoning.
Note: the dots marking the locations on the picture are about 1 mm across.

Smaller then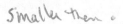

8) The Sun is much smaller than a nebula. We used a dot to represent the Sun's location in the picture. Is this dot too small, too large or just the right size to represent the size of the Sun on the picture? Explain your reasoning.

way too big since sun is too small

9) The Milky Way Galaxy is one of the largest galaxies in a group of nearby galaxies called the Local Group. The following table lists the distances to the centers of three Local Group galaxies. Draw a dot on your picture (if possible) to represent the center of each galaxy. Don't worry about the direction (left, right, up, or down) for each galaxy; just place a dot an appropriate distance from the Sun.

Galaxy	Distance from Sun (in light years)
Sagittarius Dwarf Elliptical Galaxy (SagDEG) – closest galaxy to Milky Way	80,000
Large Magellanic Cloud (LMC)	160,000
Andromeda Galaxy (M31)	2,500,000

Do any of these galaxies fit on the page? Which one(s)?

only the 1st 2, and barely

10) The objects in question 9 are all visible in the night sky from Earth. Are these objects inside or outside the Milky Way? Explain your reasoning.

Outside

11) SagDEG is approximately 11,000 ly across. Is this galaxy better represented on your diagram by a small blob or a tiny dot? Explain your reasoning, and make an appropriate sketch to represent the galaxy.

blob

12) Within the Local Group, the two largest galaxies are the Milky Way and Andromeda galaxies. From question 9, we saw that the Andromeda Galaxy was about 2,500,000 ly from us. On the picture, this spot would be 250 cm (about two and a half meter sticks) away from the dot representing the Sun.

The nearest group of galaxies to us (not counting our own Local Group) is the Virgo Cluster, about 60,000,000 ly away. How many centimeters away would this cluster be on our picture? How many meters away would this be?

6000 cm

60 m

Imagine that you have received six pictures of six different children who live near six of the closest stars to the Sun. Each picture shows a child on his or her 12th birthday. The pictures were each broadcast directly to you (using a satellite) on the day of the child's birthday. Note the abbreviation "ly" is used below to represent a light-year.

- John lives on a planet orbiting Ross 154, which is 9.5 ly from the Sun.
- Peter lives on a planet orbiting Barnard's Star, which is 6.0 ly from the Sun.
- Celeste lives on a planet orbiting Sirius, which is 8.6 ly from the Sun.
- Savannah lives on a planet orbiting Alpha Centauri, which is 4.3 ly from the Sun.
- Inga lives on a planet orbiting Epsilon Eridani, which is 10.8 ly from the Sun.
- Ron lives on a planet orbiting Procyon, which is 11.4 ly from the Sun.

1) Describe in detail what a light-year is. Is it an interval of time, a measure of length or an indication of speed? It can only be one of these quantities.

2) Which child lives closest to the Sun? How far away does he or she live?

3) What was the greatest amount of time that it took for any one of the pictures to travel from the child to you?

4) If each child was 12 years old when he or she sent his or her picture to you, how old was each of the children when you received their picture?

   John _____   Peter _____   Celeste _____

   Savannah _____   Inga _____   Ron _____

5) Is there a relationship between the current age of each child and his or her distance away from Earth? If so, describe this relationship.

6) Imagine that the six pictures were broadcast by satellite to you and that they all arrived at exactly the same time. For this to be true, does that mean that all of the children sent their pictures at the same time? If not, which child sent his or her picture first and which child sent his or her picture last?

(margin, handwritten, rotated) from million years #1, and Photons from R-born, #1, Photons from #1, #1, Acted out. A person each for #1, #1,

7) The telescope image at the right was taken of the Andromeda Galaxy, which is located 2.5 million ly away from us. Is this an image showing how the Andromeda Galaxy looks right now, how it looked in the past or how it will look in the future? Explain your reasoning.

Past - light just now arriving

8) Imagine that you are observing the light from a distant star that was located in a galaxy 100 million ly away from you. By analysis of the starlight received, you are able to tell that the image we see is of a 10 million year old star. You are also able to predict that the star will have a total lifetime of 50 million years, at which point it will end in a catastrophic supernova.

a) How old does the star appear to us here on Earth?

10 million yr

b) How long will it be before we receive the light from the supernova event?

40 million

c) Has the supernova already occurred? If so, when did it occur?

yes — light left 100 million years before.

40 million years after 100 million years ago 100−40

9) Imagine that you take images of two main sequence stars that have the same mass. From your observations, both stars *appear* to be the same age. Consider the following possible interpretations that could be made from your observations.

a) Both stars are the same age and the same distance from you.
b) Both stars are the same age but at different distances from you.
c) The stars are actually different ages but at the same distance from you.
d) The star that is closer to you is actually the older of the two stars.
e) The star that is farther from you is actually the older of the two stars.

How many of the five choices (a-e) are possible? Which ones? Explain your reasoning.

a,e e could be correct, but unless its far away (like in another galaxy) the "look back time" will be smaller then the uncertainty in the age.

The two drawings below represent the same group of galaxies at two different points in time during the history of the Universe.

Early Universe   Universe Some Time Later

1) Examine the distance between the galaxies labeled A – E in the *Early Universe*. Are all the galaxies the same distance from each other?

2) Describe how the Universe changed in going from the *Early Universe* to the *Universe Some Time Later*.

3) Do the galaxies appear to get bigger?

4) Will stars within a galaxy move away from one another due to the expansion of the Universe? Explain your reasoning.

5) Compare the amount that the distance between the D and C galaxies changed in comparison to the amount that the distance between the D and E galaxies changed. Which galaxy, C or E, appears to have moved farther from D?

6) If you were in the D galaxy, how would the A, B, C, and E galaxies appear to move relative to your location?

7) If you were in the D galaxy, would the A, B, C, and E galaxies all appear to move by the same amount in the time interval from the *Early Universe* to the *Universe Some Time Later*?

8) Imagine that you are still in galaxy D. Rank the A, B, C, and E galaxies, in terms of their relative speeds away from you, from fastest to slowest.

9) Now imagine that you are in the E galaxy. Rank the A, B, C, and D galaxies, in terms of their relative speeds away from you, from greatest to smallest.

10) Is there a relationship between an object's distance away from you in the Universe and its apparent speed? If so, describe this relationship.

11) Would your answer to question 10 be true in general for all locations in the Universe?

12) Consider the following discussion between two students regarding the possible location of the center of the Universe.

Student 1: *Since all the galaxies we observe are moving away from us, we must be at the center of the Universe.*

Student 2: *If our Milky Way Galaxy were like galaxy A and if the Pinwheel galaxy were like galaxy B, then we would both see all other galaxies moving away from us. So I'm not sure if our Milky Way Galaxy would be at the center or if it would be the Pinwheel galaxy.*

Do you agree or disagree with either or both of the students? Why?